Race, Rhetoric, and Technology

Searching for Higher Ground

NCTE-LEA Research Series in Literacy and Composition

Andrea A. Lunsford and Beverly J. Moss, Series Editors

Alsup • *Teacher Identity Discourses: Negotiating Personal and Professional Spaces*

Banks • *Race, Rhetoric, and Technology: Searching for Higher Ground*

The *NCTE-LEA Research Series in Literacy and Composition* publishes groundbreaking work on literacy, on composition, and on the intersections between the two.

Volumes in this series are primarily original authored or co-authored works that are theoretically significant and hold broad relevance to literacy studies, composition, and rhetoric. The series also includes occasional landmark compendiums of research. The scope of the series includes qualitative and quantitative methodologies; a range of perspectives and approaches (e.g., sociocultural, cognitive, feminist, psycholinguistic, pedagogical, critical, historical); and research on diverse populations, contexts (e.g., classrooms, school systems, families, communities) and forms of literacy (e.g., print, electronic, popular media). The intended audience is scholars, professionals, and students in a range of fields in English studies, including literacy education, language arts, composition, and rhetoric.

For additional information about the *NCTE-LEA Research Series in Literacy and Composition* and guidelines for submitting proposals visit **www.erlbaum.com** or **www.ncte.org**.

Race, Rhetoric, and Technology

Searching for Higher Ground

Adam J. Banks

Syracuse University

NEW YORK AND LONDON

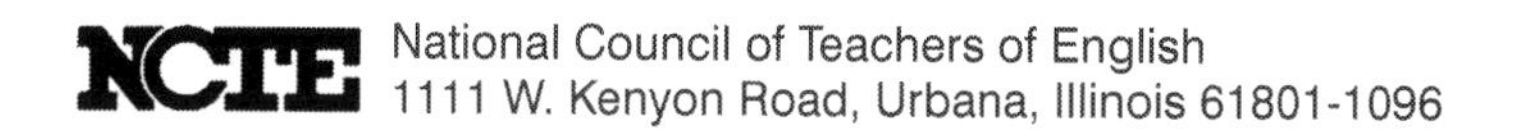

A co-publication of LEA and NCTE.

First published 2006 by Lawrence Erlbaum Associates, Inc., Publishers
10 Industrial Avenue
Mahwah, New Jersey 07430

National Council of Teachers of English
1111 West Kenyon Road
Urbana, IL 61801-1096

This edition published 2013 by Routledge

Routledge
Taylor & Francis Group
711 Third Avenue
New York, NY 10017

Routledge
Taylor & Francis Group
2 Park Square
Milton Park, Abingdon
Oxfordshire OX14 4RN

Cover design by Kathryn Houghtaling Lacey

Library of Congress Cataloging-in-Publication Data

Banks, Adam J. (Adam Joel)
Race, rhetoric, and technology : searching for higher ground / Adam J. Banks.
p. cm.
(NCTE-LEA research series in literacy and composition)
Includes bibliographical references and index.
ISBN 0-8058-5312-X (cloth : alk. paper)
ISBN 0-8058-5313-8 (pbk. : alk. paper)

1. African Americans—Social conditions—1975– 2. African Americans—Intellectual life. 3. African Americans—Communication. 4. Digital divide—United States. 5. Technology—Social aspects—United States. 6. Rhetoric—Social aspects—United States. 7. United States—Race relations. 8. Racism—United States. I. Title. II. Series.
E185.86.B264 2005
305.896'073—dc22

2005040144
CIP

To Stevie, for continuing to call us to find higher ground.

To all those old souls who made SOUL a style,
a mode of resistance and participation,
a vision of the best we can be.

To my parents for making me a midnight believer.

To KG for giving me an academy worth believing in.

Contents

Foreword

Out. That place of performance excellence beyond the ordinary, as in "she or he just took it out." Not to a totally strange place, but one familiar enough that we could have imagined it or even arrived at it ourselves had we been so astute, symbolic action close enough to what we already know that we can readily recognize the genius involved. This is my initial response to reading *Race, Rhetoric, and Technology: Searching for Higher Ground*. With this debut book, Adam Banks has taken it, yes, higher, but also out, has taken the study of African American Rhetoric, perhaps even the field of Rhetoric and Composition more broadly, fully into the twenty-first century, for this is the most sweeping, intelligent, and provocative articulation we have of racialized rhetoric and social justice, and their connections to issues of technology and design.

As I draft this present text, I am also watching *FDR: A Presidency Revealed* and learning how Roosevelt worked painstakingly—and quite painfully—to construct a visual rhetoric that projected him to the general public as an able-bodied man, a design strategy, a technology given the mechanical devices needed to complete the illusion, that enabled him to become the most important politician of the twentieth century. Whatever memorable words President Roosevelt spoke were made possible by his successful ruse. As I refocus on Banks' manuscript, I am reminded (television's direct influence on my education) of how cogent his move is to treat Martin Luther King, Jr. and Malcolm X, the two most-studied African American rhetors of the twentieth century, in a similar fashion. He clearly links their awareness of technological possibilities to the brilliance of their rhetorical strategies. Overall, with his emphasis on things technological, Banks challenges the dominant, logocentric or "great-speaker" constructions of the African American rhetorical tradition, but he has done so in a way that demonstrates clear respect for the "word-based" scholarly predecessors (myself included) who are objects of his critique. He does not argue that his elders got it wrong, simply that by ignoring technology they have missed an opportunity to present a fuller picture. This line of reasoning is certainly fair. But there is much more at stake with Banks' project than a precocious intellect trying to get props. He argues forcefully that to the extent that technology structures opportunity in the United States, racism and racialized exclusions are technological issues. It follows, for him, that if our nation ultimately fails to eliminate racist exploitation from its "grand technological experiment," then it cannot fulfill its democratic promise. A primary aim of his analysis, therefore, is to set an agenda for discussion and action.

Drawing upon his own educational journey, Banks highlights his understanding of the Digital Divide, which he views as an information-age manifestation of the Racial Ravine. Perhaps it's a good thing our author didn't get that Atari 2600 he coveted as a child; he would have lost his critical edge. Fortunately, we have him around to suggest the development of a "Black digital ethos," a multi-faceted mindset with which to assess appropriately the discourse on technology, the efforts at technology activism, and a variety of educational proposals. In short, computers alone are never a panacea. We have to contextualize the use of computers to make helpful judgments about their efficacy in terms of the educational development of African Americans. The game continually changes. "Keep this Nigger-Boy Running" read the haunting note in Ellison's *Invisible Man*. Many in positions of power and authority today, whether consciously or not, honor that directive to keep Black folks from getting grounded in their own best interests. Only an informed opposition has a chance to counter.

This is not to argue against access to computers, just that in racialized America access is always racialized to some extent. Today's *New York Post* features an article titled "Spare E-Dime?" The story covers the phenomenon of cyberbegging. Privileged folks, mostly White I believe, have been hustling up tuition or wedding expenses. One woman raised 20K over the Internet to pay off credit-card debt. Now, an old New York boy myself, I don't advocate cyberbegging as a comprehensive social program for so-called minorities in the city, but I do wonder how many sandwiches New York subway panhandlers, disproportionately so-called minorities, could purchase with 20K.

But perhaps I digress. The thing to do is to speed on to a reading of Adam Banks. Note his technology determinism, but pay closer attention when he indicates capital's relationship to technological developments. Read the plight of African Americans as a metaphor, if you need to, for other disempowered groups. But mainly read. The message is right. The flow is tight. And don't take all night.

—Keith Gilyard
April 17, 2005
University Park, PA

Preface

The year 2005 presents African Americans with a series of stark realities. Fifty years after both the landmark Brown decision and Montgomery Bus Boycotts; 40 years after the assassination of Malcolm X and after Lyndon Johnson famously sighed "there goes the South for a generation" after the passage of the Civil Rights Act; 30 years after a second round of White flight inspired by court orders requiring school districts to comply with Brown and a labor market casually destroyed by automation and global capitalism left urban centers all over the country emaciated and Black communities obliterated; 20 years since the crack epidemic and AIDS took the ridiculous nature of Black pain and made it sublime; and 10 years since the emergence of the Digital Divide as a concept that makes the connections between Black economic, political, social, medical, and psychic pain and the uniquely American commitment to technological advancement more obvious than they have been since the days of slave labor and early industrialization, it is clear that the faith that the dark past has taught us is being sorely tested. It is also maddeningly clear that technologies and the roles they play in American society are still deeply implicated in the stubbornness of American racism and the legacies of exclusion rooted in that racism that continue to test us.

The questions remain, hauntingly: Is it possible to make this nation a just one for Black people? Is it worth the struggle? Can technologies really be used to serve liberatory ends, or is that hope just another pipe dream calling us upon our awakening to resist! resist! resist! the shiny boxes we're all sold and the promise they once held?

In this book I use the concept of the Digital Divide that has been more or less present in public discourse for the last 10 years as a metonym for America's larger racial divide in an attempt to figure out what meaningful access to technologies and the larger American society can or should mean. I also argue that African-American rhetorical traditions—the traditions of struggle for justice and equitable participation in American society—have much to offer our efforts to find higher ground and understand the messiness of what we seem to have committed ourselves to in this digital age. African-American culture and freedom struggle exhibit nuanced ways of understanding the difficulties inherent in the attempt to navigate our way through the seemingly impossible contradictions of gaining meaningful access to technological systems with the good they seem to make possible and resisting the exploitative impulses that those systems always seem to present.

I argue here that African Americans must face the current problems of technology access, no matter how far removed they seem from the day to day struggles that feel so much more immediate, and that our best way through this digital madness is to remain true to those rhetorical traditions at the root of that long struggle for a more just America, even as we revise them and find new ways to apply their lessons in this new millennium. Our only hope for a technology access worth having, for a nation worth living in, lies in our renewed passion for that struggle, our renewed commitment to the jeremiadic vision of a nation and its technologies transformed by the dogged insistence that they live up to the nobility in the ideals that made those dry bits and wires and bones walk in the first place. Just as rhetors throughout those traditions have always argued that race and our ability to solve the present problems and continuing legacies of racism are the test of our democratic experiment, eliminating racism and racialized exclusions are also the test of our grand technological experiment. As I examine moments in these traditions of appeals, warnings, demands, and debates to make the connections between technological issues and African Americans' equal and just participation in the society explicit, I attempt to show that the big questions we must ask of our technologies and answer for them are exactly the same questions leaders and lay people, from Martin and Malcolm to slave quilters to Critical Race Theorists to pseudonymous chatters across cyberspace and legions more have been asking all along.

Finally, just as this book is a call for a new orientation among those who study and profess African American rhetoric, it is also a call for those in the fields that makeup mainstream English Studies to change their perspectives as well. Technology access and our ability to address questions of race and racism are the central ethical questions of our field as far as I am concerned. I try, therefore, to use this book to imagine what writing instruction, technology theory, literacy instruction, and rhetorical education can look like for all of us in a new century.

Acknowledgments

No matter how we try, these sections in books often end up sounding like clichéd awards show speeches: I'd like to thank my producer, mom and dad, and the assorted others who gave us that critical word of challenge or support along the way. And after enough, we get played off the stage, or the page, by the orchestra still feeling we haven't done nearly enough to show the depths of gratitude we feel, how central that encouraging word was in helping us figure out who we be academically, intellectually, ethically. And then there's the task of thanking those who helped us through this specific project through all of the revisions that helped turn it from the "please listen to my demo" of a dissertation to a much different, hopefully much fuller album, a much richer book.

Andrea Lunsford and Beverly Moss, my series editors at NCTE, the excitement you conveyed for this project from the beginning was amazing. Thanks as well to the staffs at NCTE and LEA for your patience and guidance as I worked through this process for the first time. My colleagues here at Syracuse University, especially Eileen Schell and Gwen Pough, thanks for your thoughtful responses to sections of this project in progress. My man, the original mack, Arthur Flowers, you're a straight treasure, in Syracuse, in the lit game, in ongoing liberation work. You still don't get the ink you deserve, but I know it's comin. Madd, madd thanks for taking me under those wide wings you got. Members of my CCR 601 class, for reading the preface and offering generous and insightful feedback. Elisa Norris heard several portions of chapters in process, listened, read, and responded thoughtfully and critically as well.

Any craftsperson worth anything, any artist worth anything has to go through the woodshed—much love, respect, and admiration for the brilliant souls who were that woodshed for me at Penn State University. My chair and mentor, Keith Gilyard—if Bill Russell the player/coach had Erving's flavor, Oscar's ability to take over, the Iceman's finger roll, and Darryl Dawkins' backboard breaking thunder, he might have been close to what you bring the academy. You let a playa handle the rock and you always coached the game, gave me the support to get through and the challenge to get over. You made that barbershop your office became, that woodshed, real, putting more Black minds out in the academy in a shorter time than anybody I've ever seen. You meant everything to my getting out here and give a hell of an example of what intellectual work can be. I can't thank you enough for letting me get on the roster.

My sistren, my shero, E'layne. Baaad linguist, scholar, bangin blues, jazz, r&b "sanga," writer, committed soul, Harriet Tubman to a brotha when I was actin like I wasn't ready to break out from UpSouth. You had my back, saw the best in me when I wasn't at my best, and kept building me even when my name might as well have been caintgitright. Ain't no story in the academy as powerful as yours, and I know because I was there almost from the beginning.

Bernard Bell, the agency, authenticity and authority. Nuff said—especially the authority. Always helping us youngins understand the importance of traditions, even when we rebel against them. Your intellectual generosity was always boss, even when we didn't see it.

Stuart Selber, Jack Selzer, Cheryl Glenn, and the faculties of the rhetoric and composition program at Penn State, and Jim Porter during that short stay at Case Western Reserve, much thanks for being wonderful guides to the discipline, for always bringing all the complexities waiting for us at the intersections of our theorizing and our practices, and for shining a constant spotlight on the messiness of it all, for challenging us to make that messiness the space of engagement. James Stewart, the young elder, there is nobody out there in Africana studies with the range, passion, intellect, and commitment you bring; Deborah Atwater, you are the one who led me rhetorically to Stevie Wonder, and to a lot of my larger thoughts about what the youngins wrongly call old-school; so much thanks to both of you for hearing the scraps of ideas as I stitched them together, and for handing me interdisciplinary thread for the task.

To my partners in crime from Penn State, much love and respect; Vorris Nunley and Howard Rambsy, the other heads in the chairs at the Barbershop, y'all are the sharpest cats I've ever seen in the classroom, always ready to mix it up, always committed to intellectual work and to Black people. To the many brilliant minds I shared classes and hallways and cubicles and bull sessions with: Aesha Adams (keep bringin the thunder, girl—you the TRUTH), Stephen Schneider, Jeremiah Dyehouse, Jodie Nicotra, Les Knotts, Celnisha Dangerfield (even Lauryn ain't got the game and brain you got): I couldn't have been surrounded by a better crew of people to learn with and from even if we weren't stuck in the hills of central PA.

Several dear friends have encouraged or endured important parts of this project too, from instant messenger conversations to long phone calls to scraps scrawled on napkins in seedy establishments. Thomas Sayers Ellis, from Pfunk to GoGo back to went went and up to where next, a hell of a mind and an open soul. Your insistence on the value of the P made me really take up the One as a model of unity in the midst of chaos. Carmen Kynard, sistafren and more, you listened to every rant, always supported, always encouraged, even when the madness in your own life reached higher and higher. Daniel Gray Kontar and Rafeeq Washington, we didn't get to kick it enough in the CLE, but your influence was always felt. Keep pennin. Ange-Marie Hancock, the baad, brilliant political scientist, for fellowship over tunafish sammiches, you're all that and I'm glad I got to see your intellect in action.

Finally, to the community of teachers, scholars, advocates, activists and everybody else who supported me long before I thought about a career as an academic, from Bobbie L Watkins and Coland Leavens, teachers of mine at Patrick Henry Junior High School in Cleveland, two of the most gifted teachers I've ever seen, teachers who kept loving, kept challenging, kept teaching, kept changing lives no matter what hell went on around them. To my Cleveland State family: Donna Whyte, Ma Washington, John R. Walton, Stanley Anderson, Stanley Gordon, Lynn Furman, Pam Charity, Adrienne Gosselin, Mwatabu Okantah, Raymond Winbush, Austin Allen, Dianne Dillard, and the memories of Wanda Coleman, Frank Adams, and Curtis Wilson. Cleveland State was (and can be again) something special if the rest of the university realizes the gifts they had and have in you. Thanks to all of you for seeing something in me worth using those gifts on.

My situation is no different than those Grammy winning artists. I've been shaped, guided, helped, picked up, and dusted off by so many people in different ways I can't call them all. To those I shout out here and those I missed, I do hope this book honors you and all you represent.

—AJB
Fall 2004

Prologue

CAN THESE DRY BONES WALK AGAIN?

I could open this book with the smug conviction that I knew there was a technological divide almost 25 years ago. My first conscious thoughts about anything technological came to the surface when I was in the fifth grade, during Jimmy Carter's second presidential campaign in 1980. My father frequently noted that if my siblings and I didn't take school seriously, we wouldn't even be able to get jobs as gas station attendants, because computers were taking over everything. That's not how I knew there was a divide, of course. I knew because Darryl and Darnell Howard, part of our football and hide and seek and Kung Fu movies crew of those years we all romanticize, had an Atari 2600, and damn it, I wanted one. I never did get one of those game machines that had us fellas all entranced, running up to their apartment whenever Ms. Howard would let them have company, glad to watch our friends play and maybe occasionally get a shot at Pac Man or any of the other two or three games I know now they ridiculously overpaid for. Mom and Pops weren't havin it—"even if we did have the money, there's no way in hell I'd spend it on that" was the familiar refrain. And hell, somewhere in me I knew it too—I was trying to graduate past supermarket sneakers (we called them Buddies back then: "Buddies, make ya feet feel fine, Buddies, cost a dolla ninety nine. Buddies, make ya feet feel great, Buddies, cost a dolla ninety eight") when you really got down to it. Fortunately, in post-riot Hough where the goal of the Snaps and Dozens games wasn't necessarily to win, but to escape being seen as the poorest, scufflinest kids in a neighborhood where everybody was poor and scuffling, I had enough Snaps and Dozens (for a shy boy) to keep the fellas off my back and get enough respect to hang with the bigger kids in those Atari and Kung Fu flick and football sessions.

So yeah, it was all about games and toys at that point, just like it is for young people today whose first introductions to things digital is often through Playstation 2 and Xbox. A couple of incidents later would add some more significant elements to my playfully nostalgic reminiscences of what "Computers" came to mean. As a seventh-grader at Patrick Henry Junior High School in 1982, where those of us tracked as "Major Work" students took pre-Algebra from Mr. Mueller, a pleasant, thoroughly giving late forties/ early fifties guy we liked but did all we could not to crack jokes on because he lived some kind of Transcendentalist life out in the woods somewhere—I think we even imagined him living in either a cabin or a tent—about 8 or 10 of us hung out in his classroom afterschool and during lunch periods whenever we could because he had hooked and crooked money from somewhere to get a computer and tried to teach us how to do things on it. Yes, it was that vague for many of us. It was a gargantuan thing, impossible to even describe now for people used to PDAs and inch-long memory sticks. Mr. Mueller had to wheel it in on a cart taller than any of us, and him too. It's hard to even remember what the processing unit looked like, but it sat on the bottom, and was connected to a TV-like monitor bigger than any television my family ever owned, and had a keyboard attached to it. There was just that one machine, and it was his pride and joy. The rest of us took a certain pride in it too, because we were the only ones in the school who got to see it, play on it, try things with it. What's hilarious for me as I remember it, and sad too, once I came to know more about the history and development of the personal computer and its use in education, is that all we ever did on it was BASIC, and the most basic uses of it. Of the ten of us, there were about two who really seemed to get it, and were doing something approaching programming on the computer that showed the results of that work in green blips across a darker, still greenish screen. But for the rest of us, the other eight or so of the ten in a school of about a thousand students who were the beneficiaries of all of Mr. Mueller's work and insistence that technology was crucial to our futures, was something like this:

```
10 print "Adam is the man";

20 goto 10

run
```

and we'd be almost as pleased as we were with Pac Man and Ms. Pac Man with the fact that the screen would fill up with something like this, if we remembered the trick of putting a semicolon at the right place in our "program:"

Adam is the man Adam is the man Adam is the man Adam is the man
Adam is the man Adam is the man Adam is the man Adam is the man
Adam is the man Adam is the man Adam is the man Adam is the man
Adam is the man Adam is the man Adam is the m

Otherwise, it would just be a boring, single-columned scroll down the screen:

Adam is the man

Adam is the man

Adam is the man

Adam is the man

Adam is the man

Adam is the man

Adam is the man

Now, even as a 11 year old seventh grader, I could appreciate all the sacrifice that Mr. Mueller put into his teaching and his attempts to get us to value computers, but that single column wasn't enough to motivate me to get to the promised land that the Bell brothers made it to as the two "experts" in BASIC at our school. I kind of envied them, but I knew I could hang with them in math, so I was straight, and besides, Ms. Watkins, the fine English teacher from Mississippi, took an interest in me as a kind of son, saw through my stubbornness and melancholy and smart mouth and a whole bunch of other stuff and prepped me for the Spelling Bee, where I won the city crown. I still hung out with Mr. Mueller, but he couldn't compete with that!

Fast forward a few years. Twelfth grade at John Hay High School, walking distance from that same post-riot Hough neighborhood I grew up in, and a stone's throw from Case Western Reserve University, where we all hung out and got thrown out after school. AP courses in English and Calculus; retaking the Algebra II/Trig course I refused to attend in the tenth grade when I was subject to some old cantankerous idiot we were all sure hated the Black students who were bused to West Tech. But he was almost fair in his scorn—he seemed to hate everyone else almost as much. So anyway, at John Hay, needing to add at least one more class, I was encouraged to take drafting with Mr. Moore, because I had designs on being an engineer. As much as I loved Mr. Moore, who was as committed and intellectual as any teacher I ever had, a brotha wasn't sitting in some Industrial Arts class. Wasn't even any cuties up in there to look at! I couldn't draw worth a lick anyway, and wasn't tryin to. So I visit the guidance counselor, try to drop the class so I could get out of school early, and end up with a Computer Science class instead. I couldn't complain much—I was actually with it, thinking there might be something interesting going on there, even though it was the very last period of the day, messing up all of my plans to be done by 12:30. And when I get to class I find myself in a room of computers, half of which worked, all of which were ridiculously old, and staring at me as the course text: a book titled BASIC. And more 10 20 30 40 .... run. I didn't even give it enough of a chance to see what BASIC might really be. I was out. Back to the counselor's office, and stuck in Geography, a class evenly split between ninth and twelfth graders repeating lessons I felt like I had in general conversations in the fifth and sixth grades when we were talking about the electoral map and John Anderson's third party chances and the Soviets in Afghanistan.

The point of the story is that, in spite of the fact that I never took to the computers I had exposure to, it took the extraordinary commitment and even visionary work of a couple of my former teachers to get technologies at least in the conversation in cash-strapped schools in a district abandoned economically and politically in a busing aftermath only Boston and Chicago could compete with. And even then it was only BASIC. I didn't realize how important that work was and how powerful that instruction, even in BASIC, could have been until much later, when in 1995, the year the Digital Divide emerged as a term coined by Larry Irving and his colleagues in the Clinton administration as well as the year William Mitchell's cyberfantasy *City of Bits* hit bookshelves, I visited my old high school. I visited the school frequently, so I don't know why I failed to notice this before, but the computer lab that I scorned was no longer there. Nor had it been replaced by anything. The digital infrastructure for all the schools' records and administrative functions was a VT 200 that seemed at least 15 years old and made me wonder how report cards ever got out. Teachers had absolutely no access to computers, the Internet, or anything else. One teacher was attempting to get a computer lab of any kind in the school, but when that lab was purchased some years later, the "business education" (read, clerical education) classes had priority. The school's library finally got four computers and dialup Internet connections in 1999, and got a second, unconnected lab installed around the same time.

There is much that this condensed sketch of my recollections of technologies in my old schools misses, and it might seem to make absolutely no connection to the initial thoughts about computer games with which I began. Let me try to connect them with another reflection: while I came of age during the revolution in computerized gaming and did all kinds of slow grinds to Roger Troutman and Zapp's 1986 song "Computer Love," and heard Prince's 1983 lament "Computer Blue" and heard funkmaster George Clinton achieve all kinds of electronic effects in his music, from "Atomic Dog" to "Computer Games," it took me until much, much later to develop anything approaching what we might call digital literacy. Let me try to make this plainer: any 6 year old who has ever touched a PlayStation I or II could beat me in the football game Madden2004 or 2003 or 2002 or almost any year, even though I like the game and would love to get better at it. Why is it that almost any 6 year old could easily whip me, a PhD (yes, in English, but still ...) who grew up in the age of computer games and who writes about technology access? The development of sports video games is the perfect example of the larger point I want to make about the utter seriousness of technology access. The ever popular John Madden-endorsed football game, like almost all other home video games, goes through changes each year. The virtual game becomes more "realistic" in each instance, with ever-increasing numbers of "plays" one can assign an offense or defense, ever-expanding controls the gamer has over the players on the field, ever more complex codes and patterns of implicit knowledge carried over from older versions that make me more confused every time I even watch my younger brothers or nieces or nephews play. Even though none of us has played organized football, my family members and friends have a far more sophisticated knowledge of football play calling and execution than I will ever have—so much so that I bought myself that long-desired Atari 2600 and its RealSports Football last year and now delight in the easy access it offers me, with its blips moving across my television screen and simple metaphors of running, passing, tackling (and especially with its uncomplicated joystick configuration).

It is clear to many by this point that my dreamy recollections of Ataris and my narrative about family members' expertise with current video games and their platforms is a parable of sorts to make sense of what has happened in the educational system and the nation at large since computers and information technologies came to dominate our understandings of education and the workplace. For all the time and effort it would take me to become competitive with my friends and family members in John Madden's namesake football game or NBA Live or any other game they enjoy, the games and maybe even the gaming systems will have changed, making my work that much harder—all of this in a situation where I can at least afford a Playstation II or Xbox if I wanted one. Take this same situation, but instead of a family's fun on some weekend when I visit home and bragging rights being the only things at stake, and the fact that I can at least choose to make all of the adjustments I would need to if I wanted to "compete," and instead imagine one where an entire group of people have been systematically denied the tools, the literacies, the experiences, the codes and assumptions behind the design choices, the chance to influence future designs and uses, and make the stakes that people's educational success, employability and thus their incomes, roles in the society, and their political power, and tie all of that to longstanding lies about that people's educability through regular news stories about their violence and failing schools and connect that to a centuries old history of outright exclusion from any education involving any technology supported by violence, terror, politics, and the definitions encoded into our nation's founding documents, and then one might understand what is at stake for African Americans with the Digital Divide.

Not only are Black people forced to catch up to technological tools and systems and educational systems to which they have been denied access, but they are required to do so in a nation (or system) in which the struggle they endure to gain any such access to any new technology, any acquisition of any new literacies, is rewarded by a change in the dominant technological systems and the literacies used to facilitate access to them, and thus the same struggle over and over again. The only difference is that the consequences are people's life chances, their material realities, whether they get to eat, have homes, and live fulfilling lives.

Think about it this way: in 1995, when the term Digital Divide began to emerge in the national conversation to signify the systematic differences in access to computers and the Internet, Cleveland Public Schools and many districts like it were still trying to get students BASIC and "basic" word-processing skills, and were struggling to get even those "basic" goals financed because of the landscape of educational funding that still has not been repaired because everyone who could afford it fled urban centers rather than have their students "forced," as they often put it, to attend schools outside their neighborhoods with children of different racial and cultural backgrounds.

During what was frequently called a digital and information revolution that leaders like Robert Reich argued would fundamentally change the nature of work and would force us to change our approaches to education, what kinds of computers and information access did school districts in Cleveland, Detroit, Philadelphia, Blair, South Carolina, Tallahatchie, Mississippi, and Los Angeles have, to prepare students for that revolution, if any? Forced to respond to a reality that a combination of information and technology access and literacy would determine students' futures, teachers and administrators went about the business of getting computers and Internet connections and training. But while they were doing that, word-processing tools like WordPerfect faded in use, Microsoft's dominance emerged even more forcefully than it had to date, and tools like Microsoft Word and PowerPoint and database technologies had gone through several new versions, each with different capabilities, different expectations of users, making real the high-stakes version of my electronic football parable. Even more significant than the changes in specific applications, the Internet emerged in the public consciousness as it changed to a graphical interface. What computers were and how they were to be used changed, the role of technologies in writing changed, and writing and literacies themselves changed. All of these changes required new skills, new processes, new behaviors, and vast investments in new equipment. But when those teachers and administrators attempted to turn over the equipment and develop the skills and the literacies and perform the behaviors and acquire the understandings, the initial impetus they had because of the currency of the explanatory power (read its ability to allow administrators and teachers and political leaders to demand funding and training and other assistance) of the Digital Divide is gone. Gone because, to most people, the Divide itself is gone.

This phenomenon is no different than what African Americans faced when they looked for ways to gain access to the agricultural technologies that governed the economic life of the nation after slavery. By the time African Americans had any semblance of a chance to learn the technological and literacy skills they needed to enter this economy, its base had shifted from agricultural to industrial. By the time labor activists and leftists like Harry Haywood and Claudia Jones and A. Philip Randolph and civil rights leaders and Black Power activists were able to shame, cajole, threaten and force the nation to do anything significant to ensure African American access to industrialized labor markets and technologies, the information age and computers emerged and then dominated. And while the information age and computers came to dominate, thousands of young Black students were encouraged, like I was, like Malcolm X was, to take Industrial Arts courses.

So high you can't get over it, so low you can't get under it, so wide you can't get around it. Like parishioners in Black churches sing about the love of their God, and like George Clinton and the Pfunk All Stars sing about the "funk," American racism and the racial exclusions that are made manifest in American life and economy seem impermeable, built not only on the idea of scientific and technological advancement, but on the planned obsolescence of our technologies. What this means for Black people is that the original sins of slavery and Jim Crow and the continual dehumanization that accompanied them are repeated, reinscribed into the life of the nation every time the technologies that govern its economic, social, and political structures change. This is so much the case that we are forced to ask of technologies, just as we have always asked of the nation: Is it possible to make this thing work, or do we resist this entire system with all of its built in exclusions and find other ways to survive? Is it time to leave altogether? Does this entire nation have to be rebuilt, redesigned, reengineered if African Americans and other people are to have any chance at equal participation in it?

I remain convinced that Black Jeremiahs, from David Walker and Frederick Douglass, were right. The stakes are high and the task is just as urgent as it was in 1830, 1877, 1909, 1965, and 1980. This nation will have to be transformed if true equality and justice are to mean anything for African Americans. In spite of the stubbornness and spitefulness of racial injustice in the Unites States, I also remain convinced that it is somehow possible to make these dry bones walk again. We know the limits, however, of looking to our legal system, or even the hearts and minds of citizens for that change. The redesign of a nation—especially this nation, in this moment—must begin with its technologies.

1

Introduction

LOOKING FOR UNITY IN THE MIDST OF MADNESS: TRANSFORMATIVE ACCESS AS THE ONE IN AFRICAN AMERICAN RHETORIC AND TECHNOLOGY STUDIES

In cyberspace, it is finally possible to completely and utterly disappear people of color. I have long suspected that the much vaunted "freedom" to shed the "limiting" markers of race and gender on the Internet is illusory, and that in fact it masks a more disturbing phenomenon—the whitinizing of cyberspace. The invisibility of people of color on the Net has allowed White-controlled and White-read publications like WIRED to simply elide questions of race. The irony of this invisibility is that African American critical theory provides very sophisticated tools for the analysis of cyberculture, since African American critics have been discussing the problem of multiple identities, fragmented personae, and liminality for over a hundred years. But WIRED readers and writers aren't familiar with this rich body of critical theory ... the struggle of African Americans is precisely the struggle to integrate identity and multiplicity, and the culture(s) of African Americans can surely be understood as perfect models of the "postmodern" condition, except that they predate postmodernism by hundreds of years, and thus contradict the notion that the absence of the (illusion of) a unitary self is anything new.

We don't need a "whole new set of metaphors for thinking about the unconscious." We need, as a culture, to pay attention to the theory and literature of those among us who have long been wrestling with multiplicity. There are many things about e-space which are not new. Yes, the Internet gives us more people writing, but I'm afraid that at the moment it gives us more of the same people writing. Let's see some real difference.

—Kali Tal, "The Unbearable Whiteness of Being: African American Critical Theory and Cyberculture." *Wired Magazine,* October 1996.

This project is an attempt to chart some of the ways African Americans have struggled to make real difference in a nation whose existence depends on rigorous commitments to technological advancement and exclusions based on race. African American rhetoric as read through a technological lens allows a thorough documentation of that struggle, and ways it can contribute to broader digital and rhetorical theory. It can also help us all—leaders, activists, scholars, and lay persons involved in dismantling the systemic supports for racism—reconfigure a sense of what that collective struggle might mean and how it can be taken up at such a difficult time in American history.

The overall argument I make is this: rather than answer either/or questions about whether technological advancement and dependence leads to utopia or dystopia, whether technologies overdetermine or have minimal effects on a society's development, or whether people (especially those who have been systematically excluded from both the society and its technologies) should embrace or avoid those technologies, African American history as reflected through its rhetorical production shows a group of people who consistently refused to settle for the limiting parameters set by either/or binaries. Instead African Americans have always sought "third way" answers to systematically racist exclusions, demanding full access to and participation in American society and its technologies on their own terms, *and* working to transform both the society and its technologies, to ensure that not only Black people but all Americans can participate as full partners.

The story of African Americans' pursuit of a *transformative access* can contribute much to rhetoric and technology theory by engaging both in a space beyond the narrow polemics of whether Technology is ultimately evil or wonderful, but rather develop and articulate models of the specific kinds of practices that can provide excluded members of society access to systems of power and grounds on which those systems can be challenged and ultimately changed in meaningful ways. This story can also provide a framework of African American rhetorical study that moves beyond the admiration of individual exemplars of rhetorical mastery or ideological debates about whether people or organizations were assimilationist or separatist, accomodationist or resistant, liberal or conservative, progressive, or radical. As important as the exemplary text or figure is, attention to the individual text, rhetor, or moment decoupled from the traditions and movements that make them possible leads to the assumption that the gifted Black rhetor is somehow an exception to, rather than the result of, "normal" speakers or writers. Such a different approach to African American rhetoric, I hope, can connect rhetoric with the "real" to provide bases for collective action while refusing to demand that people submit to a practical or ideological orthodoxy that, in the end, not only destroys individual identity but the possibility for collective action as well.

Making this kind of move calls up important definitional questions. For the purposes of this work, I define African American rhetoric as the set of traditions of discursive practices—verbal, visual, and electronic—used by individuals and groups of African Americans toward the ends of full participation in American society on their own terms. These traditions and practices have both public and private dimensions and embrace communicative efforts directed at African Americans and at other groups within the society: hence, directly persuasive public address and less overtly persuasive day to day performances that contribute to the creation of individual and group identities are all viable subjects of African American rhetorical study.

My understanding of African American rhetoric acknowledges and builds on the focus of the power of the spoken word and Black orators, but also attempts to open it up to all of the means employed throughout Black history—to value the uses to which rhetors have employed design, visual communication, electronic communication, and performance that are often appreciated but dealt with only tangentially.

The approach to technology that guides this project comes from a wide range of sources, but those sources come together in Martin Heidegger's definition of technology as a combination of instruments and processes, artifacts, and activities:

> [e]veryone knows the two statements that answer our question. One says: Technology is a means to an end. The other says: Technology is a human activity. The two definitions of technology belong together. For to posit ends and utilize the means to them is a human activity. The manufacture and utilization of equipment, tools, and machines, and the manufactured and used things themselves, and the needs and ends that they serve, all belong to what technology is. The whole complex of these contrivances is technology. (p. 5)

Heidegger's consideration of processes and needs opens space for the examination of critique and design, in addition to the use of particular artifacts. Patricia Sullivan and Jeannie Dauterman put it a different way in "Issues of Written Literacy and Electronic Literacy in Workplace Settings:" "We contend that technology, especially when it networks writers to other writers, is more than a mere scribal tool. It offers—at the very least—a connection to new sources of information, a site for rethinking structures" (p. viii).

In spite of my obvious "Amen" to Kali Tal's call that inspires my work, by embracing a shifting, yet thoroughly rooted African American self and arguing for what Deborah McDowell would call a "changing same" exigence for a collective, even unified group identity and struggle, I refuse to believe the postmodern hype that Black scholars and theorists have been forced into by the emergence of Theory as a canon trumping even that of Literature, which it worked to dislodge. Rather than look to the cyborg and perpetually fragmented selves and unyielding rupture and be resigned to either celebrate, lament, or endure them as simply the state of being we've all been reduced to in this moment, I choose to look for the ways Black people, through their rhetorical traditions, have worked to find their collective way home and heal what ranks among the greatest physical, material, and social ruptures experienced by any group in human history.

To put it differently, I choose Sixo of Toni Morrison's *Beloved* stealing away at night and walking interminable distances under threat of punishment or death to meet what other slaves called the "30-mile woman," the true partner who "is a friend of my mind. She gather me, man. The pieces I am, she gather them and give them back to me in all the right order" (pp. 272, 273) over the endless wandering and exploration of the nomads in technology theory and novels. I choose James Baldwin's tireless work as expatriate and both the ultimate insider and ultimate outsider of African American and American culture, in desperate attempt to save Black people and the nation over the cybertourist who feels free to "try on" any identity he/she/it pleases. I choose to celebrate the stories of radicals like Mama Freeman and Papa Freeman in my hometown Cleveland and many others that Robin D. G. Kelley (2002) tells in his book, *Freedom Dreams: The Black Radical Imagination,* stories of those free people who struggled for a short time and for a lifetime who dared imagine a new life, new society, new identity. I choose to honor people who labored under the constant threats of surveillance, suppression, poverty, ostracism, and death instead of those post-everything navel-gazers who remain content to point out the supposed theoretical shortcomings of people and movements; content to debate endless complexities never connected to a project to improve anything in the real or virtual worlds. Finally, again rolling with Robin D.G. and poet Thomas Sayers Ellis, I choose the Mothership and Uncle Jam over the cyborg—that Mothership and the Atomic Dog himself always making it a point to land and bring something back to the people, no matter what abstractions and fanciful flights throughout real and fantasy worlds George and PFunk might allow themselves. In these choices, I try to work through the truths African Americans have found and can build from, rather than submit to a theoretical orthodoxy that insists everything is a "social construction" and therefore not based in fact and therefore untrue: an orthodoxy that leaves little room to socially or physically construct anything outside of the current order of White, western, social, political, economic, and technological domination.

To talk about African American rhetoric, in this moment, however, by offering any unifying narratives about how that rhetorical production takes place is to invite challenges—some thoughtful, some not—about whether doing so is to essentialize Black experience. Some would argue that there are so many elements in individual Black people's experiences, that it is an obvious fiction to talk about *a* Black anything. There's gender; there's class; there is a large population of Black people in the United States who are not of U.S. ancestry since changes in the country's immigration laws in 1966 permitted African and Caribbean Blacks to come here. Although it is clearly important to examine Black experience in complex ways, and to account for all of the many other issues that intersect in the creation of Black identities, for those studying African American culture and documenting African American struggle to take up postmodernism's theoretical excesses and dismiss any construction of group identity as essentialist is not only suicidal, as Molefi Asante (1998) would note. It amounts to tearing down painstakingly constructed buildings and being content to stand forever in the rubble. Part of the difficulty with these excesses is that they seem to be applied more stringently to the very groups that have pressed for inclusion in the academy, in technology, in the nation. For example, America's heterogeneity and its difficulties constructing a sense of place for itself in a rapidly changing world leaves very few people to deny that there are themes and experiences that construct a framework for American identity that individuals and groups play into, with, and against.

African American rhetoric is particularly susceptible to charges of essentialism because its genesis as an area of study in the university is tied to the Black Power and Black Arts movements and the charges that those movements amounted to "Negro Thought Police" in an attempt to create a unity that could sustain political action. The problem with those charges is that they sometimes mistake the part for the whole. The so-called failures of the Black Arts and Black Power movements (and I believe these movements need far richer analysis before one moves to assess their so-called "failures") do not mean that positing any kind of Black unity is automatically problematic. It means that the grounds for arguing that unity have to be reconsidered, not the possibility of unity itself. I believe that African American rhetorical history shows powerful unities of identity and purpose across centuries, classes, genders, and ideologies, once we realize that unities are not absolute.

To say that one must play fuller, richer chords, and leave individuals more room for improvisation both with and against those chords is not to say that there is no music that can reflect their collective energy and aspirations. To continue the music metaphor, I'm looking for the ONE in Black rhetoric. Funk music, and particularly that made by Parliament/Funkadelic in the 1970s, articulated "The One" as the concept that guided the genre. The concept, basically, is this: members in the huge bands that made up PFunk could do almost whatever they wanted in most of the measure, but they always had to come back on "The One," or the first beat of the measure. That first beat was always heavily emphasized, in contrast to the two-four iambic pattern to much American music, and that first beat set the structure that members would respond to and against during widely varying solos, leads, and harmonies. African American rhetoric has embodied this concept to me because Black people and the rhetorical texts and forms they produced have always come together, one (figurative) nation within a nation, under a groove, in moments of urgency, of struggle toward that transformative access. Although there is no political or rhetorical utopia that allows all voices to be heard, African American rhetoric has, through a tradition of debate and dialectic, made sure many varying views received a hearing leading up to and following those moments of urgency. Its study now can honor those traditions and continue the struggle that has often been the source of its production.

The commitments with which I begin this project and the importance I attach to them make this work a strange calculus of polemic and observation. I work to chart traditions and examine them as much as I try to argue for the value of a particular lens for the rhetorical microscopes and telescopes used in that charting. To that end, I examine a range of texts—print, electronic, oral, visual—to show the ways specific African American rhetorical forms and practices contribute to the work of transforming America and its technologies, whether that transformation be achieved through critique, use, or design. Rather than attempt to be exhaustive, I hope this range (through its strengths and limitations) suggests all that is possible in African American rhetorical study, and that technologies indeed are not limited to artifacts alone. I also hope that the selection of texts and sites included here points to the importance of the endpoints and many places in between the continua that inform rhetorical scholarship: the public and the private, the serious and the recreational, the signal moment and the everyday performance.

Obviously, the parabola I sketch in the following chapters after analyzing these texts from this perspective is only one way of organizing African American rhetorical traditions among many possible ones. It represents only one set of texts out of legions that could be selected. This is one of what could be several downbeats to be found in African American rhetorical traditions. The more texts one takes up, the more the integral reaches from zero to infinity, obviously, the smoother the resulting parabola; the more nuanced a look at its trajectory one's attempt to chart it would provide. A different equation—or organizing story—could well produce a different-looking path even when the same moments, or texts, are examined. To switch the metaphor, this project is a quilt like those slaves and free Black people have been piecing together from the patchwork of our experiences, whatever materials could be begged, borrowed, bought, or stolen, and stitched together with the one aim of somehow pointing people to a freedom with technologies that ultimately transform those technologies, and the nation, and us. I proceed from what I believe to be the one constant that holds fixed regardless of the understandings of race, culture, or identity one might choose: that African Americans have always, since being brought to the American colonies and the United States as slaves, existed in a society that has rigorously enforced, and steadfastly refused to correct, a system of exclusions connected to race. The variables of how Black people see themselves, understand race, and want to participate in the nation and its technologies have been organized around the consistent struggle to both participate in and change them. This bidirectional look at both the systematic nature of the struggles we face and our own incredible agency and innovation in the midst of those struggles must be focus of any examination of African American life, activism, education, identity, celebration, culture in this moment.

In the second chapter, "Oakland, The Word, and The Divide: How We All Missed the Moment" I contextualize the emergence of the Digital Divide as a concept through a look at the debates and discussions about race that marked rhetoric and composition, technical communication, and African American rhetoric when the Digital Divide entered public conversation in the late 1990s. I then make the case for a technological reading of African American rhetorical history as Black people's pursuit of transformative access to the technologies that construct and comprise American life. In making that argument I examine how the "digital divide" and the attendant discussions of technology access this concept sparked provide both an opportunity and a convenient metaphor for understanding the legacy of racism and African Americans' responses to it. I contend in this chapter that the Digital Divide is a rhetorical problem at least as much as it is a "technological" one and examines Rhetoric and Composition's history of exclusion of African Americans, especially as reflected in its turn to the technological (an exclusion that takes place even as the field has actively sought more progressive ways of dealing with the legacy of race). Although this turn to the technological takes place at the exact same time as the emergence and popularization of the Digital Divide as a concept in the national lexicon, African Americans' relationships to learning and communication technologies and their contributions to Rhetoric and Composition scholarship are elided almost entirely, through a reduction of Black students and Black participation in the field to a debate over Ebonics. The irony of this elision is particularly compelling given this moment in the field's history: two decades after the Ann Arbor trial, 30 years after the Conference on College Composition and Communication's passing of the "Students' Right to Their Own Language" language policy document, and 50 years after Brown v. Board of Education and the beginnings of linguistic scholarship showing the complexity and history of African American English.

I argue here that a useful understanding of African American rhetoric must account for both public and private kinds of persuasion, of communication by experts and lay users of a system, as well as non-users who attempt to use or participate in it. Finally, I present in this chapter a taxonomy of access that shows a meaningful access, a transformative access occurs simultaneously along the connected axes of critique, use, and design. One reason I argue for the necessity of such a multidimensional view of rhetoric and technology access is my belief that one cannot afford to be forced to choose between Rhetoric and Composition, between skills and critical thinking, between technological literacy and essayistic literacy, between being a technophile or a Luddite. These are all false choices. An extension of this argument guides the rest of this project: rhetorical and technological education must take up all three of these axes in theory, pedagogy, and practice with the focus on helping students employ each and all toward access and/or transformation of the spaces they occupy as they see fit.

In chapter three, "Malcolm, Martin, and a Black Digital Ethos," I use these central figures in African American rhetorical study to demonstrate the pursuit of technological transformation at work in even the most traditional texts. This chapter presents King and Malcolm X not only as exemplars in the use of specific technologies (in this case television) toward rhetorical ends, but also as directly engaged in debate about the potentials and problems in individual technologies and our larger commitments to technological advancement. This chapter offers readings of Malcolm X's argumentative skills at work under intensely constrained and biased conditions during the interviews that were included in the television documentary "The Hate that Hate Produced." This chapter also shows the varied understandings of the electoral franchise as a technological system both Malcolm X and Martin Luther King exhibit in the classic speeches "The Ballot or the Bullet" and "Give Us the Ballot We Will Transform the South." My readings of these speeches show King and X demonstrating awareness along the entire taxonomy of access I develop in chapter two, toward the goal of transforming both how the franchise works as a technological system and the nation as a whole. I move from this discussion of a particular tool and system to Martin Luther King, Jr's grasp of our relationships with technologies more broadly defined as the central ethical issue in America in one of his final speeches "Remaining Awake Through a Great Revolution," delivered on March 31, 1968, just days before he died. I compare King's understandings of technological issues in this speech with arguments that technology theorist Andrew Feenberg makes throughout his work, but notably in the essay "Subersive Rationalization: Technology, Power, and Democracy" to show just how rich King's thinking on this subject was.

Chapter four, "Taking Black Technology Use Seriously: African American Discursive Traditions in the Digital Underground," is concerned with the ways users of the Web site BlackPlanet keep Black discursive traditions alive online. This chapter takes up the issue of transformative uses of rhetorics and technologies more explicitly than some of the other chapters and in a contemporary moment by looking at how users of BlackPlanet counter the "unbearable Whiteness of being" Kali Tal challenges in online culture.

This chapter raises the question of what happens to African American language and discourse patterns in written spaces, and online spaces in particular. I argue here, that despite of the prevalence of work on the oral elements of African American speech and discourse, and the dominance of early cyberspace theory that argued race and culture were irrelevant online, African American language and discourse not only live, but thrive online. The strong presence of African American language and discourse on BlackPlanet speaks not only to their richness but to the ways African Americans can make cyberspace an "underground" that counters the surveillance, censorship, and suppression that always seem to accompany Black people speaking, writing, and designing in more public spaces—spaces that often seem to say that whatever oral traditions one might bring to the classroom workplace or cyberspace, written English is, and will be, White by definition.

Chapter five, "Rewriting Racist Code: The Black Jeremiad as Countertechnology in Critical Race Theory" examines Derrick Bell's groundbreaking book *And We Are Not Saved: The Elusive Quest for Racial Justice* for its use of and engagement with the African American Jeremiad. Bell invokes the history of the Jeremiad and wrestles with that history in order to "make a way out of no way," to create a rhetorical situation for discussing the role of the law in upholding racism where in many ways none existed before Critical Race Theorists emerged as a distinct body of scholars from the Critical Legal Studies movement in the 1970s. That lack of a rhetorical situation lied in mainstream legal scholars' dismissal of Critical Race Theorists' critique of the ways that conventions of American jurisprudence and legal scholarship acted as a set of codes, programmed into the interface of the American legal system to mask and uphold American racism. Thus, I argue that the conventions of legal scholarship and jurisprudence provide an example of how language itself can be technologized in genre. Bell's engagement of the Jeremiad and its history in African American struggle leads him to an intervention in the genre of American legal scholarship that serves as a "countertechnology." This particular use of genre, through Bell's clear mastery of the conventions of mainstream legal scholarship *and* his insistent departures from them provides a model for attacking the ways power and exclusion can be encoded in rhetorical forms.

This particular kind of attack, from a position that is both thoroughly inside and thoroughly outside a system shows how Black rhetors have intervened and can intervene in systems of power that have excluded African Americans because of the assumptions that we simply "don't do" law or technology or rhetoric. It can also reinscribe the stories of those people on whose behalf activists argue back into central roles in those arguments.

In the discussion of African American quilters that makes up chapter six, "Through this Hell into Freedom: Black Architects, Slave Quilters, and an African American Rhetoric of Design" I show that African American aesthetic principles that are often reduced to the level of "style" are not merely about the ornamental but rather derive from a worldview and the same traditions of struggle that the preceding chapters begin to chart. The legacy of Black quilters demonstrates the importance of visual rhetorics and the design of texts, spaces, ritual, and technologies to African American struggle in spite of the lack of attention they have received to date. Simultaneously, however, it insists on the importance of Black women as central to the history of African American rhetoric and technology histories, regardless of the fact that both of those histories have been as rigorously exclusive of women as they have of Black people. The example of these quilters also helps to demonstrate that the brilliance of African American rhetoric and technology history is not only in the prototypical text, in the person singled out because "he's so well spoken!," but in the daily, collective work of people throughout the culture. The quilters show the power of the traditions that inform the singular, powerful voices of an Ida B. Wells or Marcus Garvey rather than the mistaken notion that those voices are somehow exceptions that prove the rule and justify the assumptions that are made about the ability and potential of millions of African Americans.

In the conclusion, "Searching for Higher Ground: Transforming Technologies, Transforming a Nation," I return to the idea of "The One," arguing that technology access is the ethical issue that should ground African American rhetoric, technical communication, technology studies, rhetoric and composition, and Africana studies. My point here is to issue a call for struggles toward a transforming access, but also to note in that call that technology access is the issue that can spark scholarly and practical dialogue within and across disciplinary lines that have to date been difficult to cross. The search for the higher ground of a transformative access to technologies can, in other words serve as that downbeat that can spark collective action on behalf of those we serve without losing the individual funk and flavor we all bring to that struggle. I also call scholars, teachers, activists, and laypersons to avoid the trap of remediation as our only tool for addressing systematic differences in access to technologies and writing spaces. I offer practical advice for teachers and scholars beginning to take up these issues in their classrooms and to students and scholars wondering what a scholarly agenda at the intersection of African American rhetoric and technology studies might look like.

Why is technology access the central issue in African Americans' continual struggle for justice and equality? Why should it be the major ethical issue in rhetoric and composition as well as in technical communication? There are many reasons for centralizing access in this way, but it comes down to this: more than mere artifacts, technologies are the spaces and processes that determine whether any group of people is able to tell its own stories on its own terms, whether people are able to agitate and advocate for policies that advance its interests, and whether that group of people has any hope of enjoying equal social, political, and economic relations. All of the labor issues that have left urban areas decimated while young Black women and men are warehoused in prisons that reward rural areas with larger census counts and more jobs are connected to our technologies. The education system that continues to dehumanize young African American students both suffers crippling gaps in mere material access to computers and digital technologies and wastes incredible sums of money on computers and software programs that continue the "miseducation of the Negro" in curricula that almost never move beyond remediation in a political discourse that still labels those students, their schools, and their families as failures. The quality of health care African Americans have access to is still at the root of differences in life expectancy between Blacks and other racial and ethnic groups from infancy to old age. Police still operate in systems that refuse to punish them for using the "maximum allowable force" to punish suspected criminals and innocent civilians even when they are already subdued. The technology sector still makes excuses for not developing and employing African American talent, not only because of the offshoring and outsourcing of jobs, but because its companies often choose instead to spend millions of dollars recruiting, transporting, and educating people from countries everywhere else in the world. Racism is enforced and maintained through our technologies and the assumptions we design and program into them—and into our uses of them. Without systematic study of our relationships with technologies and technological issues, we remain subject to those technologies and the larger patterns of racism and racial exclusion that still govern American society.

2

Oakland, The Word, and The Divide: How We All Missed The Moment

I once thought I could never imagine what it would have been like to be among the newly freed slaves after the shock of it all when the word finally came down, after the Juneteenth celebrations, after the pained attempts to locate lost family and loved ones, when the difficult work of reconstruction sat there waiting to be done. Same thing with those who helped to build Black lives and opportunities anew after the decades long struggle of African American leftists, moderates, liberals, and conservatives resulted in the temporary victories of the Civil Rights and Voting Rights Acts. It all seemed so basic when I learned bits of the histories of those moments: how could we not have gotten further? How could the rare coalescence of a national political will to change and legislation attempting to make that will some kind of reality not result in more tangible progress? How could Washington and DuBois still be debating what should have been obvious more than 30 years after emancipation? How could activists be left with such shambles of an education system 50 years after *Brown v. Board* when Blackfolk were infected with a euphoria that had them chanting "free by 63!"?

Why is it that we end up in this endless cycle when we learn about Black freedom struggle? Over and over and over again, victory, then inertia, then contentiousness then the sickening feeling that such history making victories made little difference in the long run? As an undergraduate with all of the idealism of 18- and 20-year-olds everywhere and all of the certitude that things were so clear and so easy, none of this made sense. The crack epidemic making shambles of my family and neighborhood at the exact same time Dr. King's birthday was being recognized as a national holiday? Surely the Negroes in whatever parallel universe might be watching us were shaking their heads as I was. Then I saw, on a much smaller scale, admittedly, how these travesties take place: the Digital Divide and

Oakland. What happened (and didn't happen) with the alignment of these two national discussions in rhetoric and composition's theorizing and teaching, showed me what happens when Black warriors get caught on the wrong end of time warps being forced yet again to defend Black humanity, and when both well meaning and reactionary elements of the mainstream refuse to change. This is partially out of their insistence on subjecting that very humanity to debate and partially out of a blindness to the grounds on which we're all forced to live out our collective humanity, shifting rapidly while forcing that very, very old debate, in a nation where seemingly promising legislation or policy turned out to be nearly bankrupt.

This is exactly what happened in the late 1990s—the Department of Commerce introduced the Digital Divide as a concept to acknowledge the systematic differences in technology access that African Americans, other racial minorities and those in rural areas experienced and attempted policy initiatives that members of the Clinton administration thought would help to erase those gaps. Although what seemed to be a promising bit of political will emerged for educators to address problems of technology access in their schools, colleges, and universities, this issue failed to even make it onto academics' collective radar, with few exceptions. And a major part of why educators all over the country, at all levels, missed it, is because of their inability to avoid, yet again, the debate over Black humanity—in this case, educability in standardized English, because of the Oakland controversy.

Rhetoric and composition, as well as the technology sector in American society, have functioned very much in the same way as the legal system (that I examine in chapter four), in that each rests on a history that has branded African Americans as utter outsiders, unworthy of full, equitable, and just access because they are non-technological, unable to learn Standard English, in essence, non-citizens. Because of the persistence of these constructions, access to technologies and the discursive practices that determine power relations in our society, the Digital Divide, and the larger history of African American is, essentially a rhetorical problem. And because African American exclusions from the educational system that determines access to employment (and therefore the technologies that undergird the American economy) are so rooted in the specter of the Ebonics speaker and writer, the rhetorical problems that dominate understandings of race in our discipline are technological problems.

The history of these constructions extends back into Enlightenment notions of race and forward into the present. Neither rhetoric and composition nor the technology sector have found ways to discuss their continued exclusions of Black people, both continuing to define the rhetor and the technology user as White by default. This longstanding theoretical blind spot is especially pronounced in a field like English Studies, where race, technology, and questions of access are all addressed, sometimes even energetically, but where the connections between them are almost never explored. This chapter is an exploration of that odd silence as it has been carried out in the major journals and some other writings in composition and technical communication over the last 10 years, as well as in the development of African American rhetoric since it entered the university as an area of study. I examine this problem of representation by looking at the flashpoint where the Digital Divide and the Oakland controversy collided, and I argue that while there are staggering silences on the connections between these issues, silences that were made even more profound by the missed opportunities that national conversations about the Digital Divide provided, the twin sites of composition and technical communication are especially productive sites from which to address problems of technological access. African American rhetorical study can help bridge those sites, end those silences, and bring new theoretical tools to bear on the conversations about the relationships between technologies, literacies, and discourses that English Studies more broadly has plunged itself into.

RACE AND TECHNOLOGY ACCESS DURING THE GREAT EBONICS DEBATE

It might seem entirely irrelevant to some, but I'm fascinated with the fact that during the period of what is often called the greatest advance in technology in the last 100 years—and what many still like to suggest is the most important communication technology to emerge since the printing press—the nation and English departments throughout were mired helplessly in yet another Ebonics debate we've already been through decades ago.

I'm still fascinated with the simultaneity of these events, nearly 10 years later, for many reasons: I wonder why the Oakland School Board's claim on federal resources for educating Black children caused such a furor two decades after the Ann Arbor trial, 30 years after the CCCC "Students' Right" document, and almost 50 years after the Brown decision; why, almost to a person, Black "leaders" and advocates sounded just like White conservatives (and still do, given the despair and rage in Bill Cosby's recent rants about Black children not wanting to learn Standard English and their parents not caring enough to ensure that they do); and why assumptions about the supposed inferiority of African American varieties of English that have been dismantled consistently by Black and White linguists alike over the last several decades still hold such force, especially with language teachers and professionals. The most important reason I'm still fascinated with the connection between these events is because of what wasn't happening during that debate: serious, thoughtful discussion about race and the problem of access to computers, the Internet, and information technologies.

Although African American engineers and science professionals, and even some mainstream policymakers have long understood the importance of equal access to these technologies, writing and communication teachers of all races have been mostly silent on the subject. If, however, the most important characteristic of computers and the Internet is their role as communication technologies—if, as Jay David Bolter and many others have argued, these tools have begun a revolution in communication more significant than any other in the 500 years since Gutenberg's printing press, the results for African Americans, who have consistently and often systematically been denied access to these technologies, will be catastrophic.

This said, English departments (specifically rhetoric and composition and technical communication programs) have been staggeringly silent about the problematic relationship between race and technology access that became crystallized in the term "Digital Divide." This silence is a peculiar one, given the facts that composition as a field owes its existence in many ways to the project of making equal access to higher education real, and has worked diligently, if not always effectively, at important points in its history to address the role of race in those struggles, especially as that struggle has been made manifest in scholars' and practitioners' longstanding debates about Ebonics. The silence is even more stark if one really considers technology access to be a rhetorical problem, because if it is, technical communication and rhetoric and composition more broadly are the intellectual spaces within English Studies, and maybe even the university as a whole that have the potential to do the most to address it. I use the rest of this chapter to examine that silence over the last decade as it echoes throughout journals and books published in composition and technical communication, as well as in African American rhetoric to show just how little has been said, offer an assessment of those attempts scholars have made, and identify places where we can speak across the divides inside our own discipline that maintain the severity of the Digital Divide that remains so pervasive. The point of this rehearsal of the work that has been done is not to castigate individual scholars or the allegiances that exist within our discipline. Rather, it is to show the opportunity we all missed but can still grasp, in adding depth to the shallow treatment technology access received when it finally did attract the attention of policymakers in the Clinton and Bush administrations; to show just how promising extended dialogue across our own disciplinary lines could be in making real progress in improving technology access for African Americans and all Americans, in making inquiry into technological issues a real part of the intellectual work of rhetoric and composition; and possibly—if genuine dialogue across those disciplinary silences ever begins—reshape the futures of rhetoric/composition, technical communication, and African American rhetorics.

During the last ten years since the Digital Divide and broader technology access issues emerged in the national conversation, not a single article in the three major technical communication journals, *Technical Communication, Technical Communication Quarterly,* and the *Journal of Technical Writing and Communication* addresses the Divide or any technology access issue, although all of these journals frequently take up questions of the promises and perils of computers and the Internet. No article in *Technical Communication* addresses any issue related to race and technology—in fact, no article in the journal takes up the question of race at all—in spite of a significant presence of African-American and Latino/a engineers (because engineering and science students are the major audience for many technical communication programs and courses) in the workplace. When "cultural" issues are raised, the subject is broached in the service of global capitalism, focused on international students, employees, or clients. International in these cases often means those from Arab, Asian, or European countries. *Technical Communication Quarterly* manages a grudging nod in the direction of Black people with Heather Brodie Graves and Robert Graves' contribution on the need for cultural sensitivity in technical editing, "Masters, Slaves, and Infant Mortality." The *Journal of Technical Writing and Communication* continues the silence on access, but includes two articles on African Americans and communication: a study of Ebonics that purports to examine student attitudes toward the language variety by "translating" a Jesse Jackson speech into Ebonics. This article says nothing about the specific features of African American varieties of English their translation used, however, or why their version focused on the features it did. The authors of the second article present it as "a descriptive study of *the* Black communication style by African Americans within an organization." This 1997 article, "A Descriptive Study of the Use of *the* Black Communication Style by African Americans Within an Organization," by Vonnie Corsini and Christine Fagliasso, ends with the familiar call for an awareness of cultural differences.

Composition hasn't done much better. The discipline's major journal, *College Composition and Communication* has published very few articles on technology access and writing instruction: Cynthia Selfe's 1997 chair's address to the annual conference, "Technology and Literacy: A Story about the Perils of Not Paying Attention," and "The Politics of the Interface: Power and its Exercise in Electronic Contact Zones," a 1994 article by Cindy and Richard Selfe demonstrating that computer and other technology interfaces can uphold the very exclusions many thought they would eradicate. The good news is that their work has helped to push composition in a more progressive direction than technical communication, one that acknowledges the political and social forces at work in the development and use of any technology—and acknowledges racism as one of those forces. Unfortunately, that's about it.

How is it that the convergence of race and technology in the problem of differentiated access is ignored, evaded, or elided in a discipline that has long struggled with issues of race and access, and one that sees its work as teaching students and professionals to become critical users of technologies? How is it that the ways African Americans and other people of color use digital technologies are entirely neglected even in the few conversations that do take place? This discursive divide occurs in part because both areas have simply been slow to make the connection. Charles Moran calls scholars to this theoretical nexus in his essay "Access: The 'A' Word in Technology Studies," arguing that composition has to carefully, thoughtfully engage problems of race, gender, and wealth differences and what their resulting problems of access portend for teaching, research, and writing. This article appears in 1999, almost 2 decades after the emergence of the journal, *Computers and Composition,* and almost 5 years since the Department of Commerce's landmark Digital Divide reports, with Moran acknowledging that his work represents first steps for this kind of inquiry: "most of us simply do not deal with the relationship between wealth and access. I think of some of the major texts in our field ... none of which raises the question of access in a substantial way" (p. 210). Moran continues, with a reflection on the institutional side of this neglect, "I would add to this list too, university alumni magazines and public relations documents that boast of their institution's technology without mentioning the fact that it is available only to a privileged few" (p. 210), and adds that scholarly anthologies in composition are complicit as well. Moran's critique of university public relations organs might seem trivial compared with the curricular and infrastructure work that does not happen, or universities' policy failures and their inability to reach out to communities and school districts that are desperate for their leadership and support, but his point is that institutional chest thumping to reach out to those who have been privileged while those failures remain is hypocritical, especially when many such guides and newsletters and magazines often find women and people of color to pose for token pictures to tout their commitments to "diversity."

James Porter does important work to clear out space for issues of access in the discipline in his book *Rhetorical Ethics and Internetworked Writing* that starts to take into account social and pedagogical issues, even if the book seems, at first glance, not to directly address how race and ethnicity might affect the definition of access he offers. Porter notes that access "may well be the number one ethical issue for internetworked writing" (p. 102), citing a need for scholars to move beyond the focus of most conversations about technology and access on the "haves" and find ways to genuinely include the "have nots." Cynthia Selfe also calls attention to the importance of access, and explores the importance of race in that conversation in her book-length treatment, *Technology and Literacy in the Twenty First Century: The Importance of Paying Attention*:

> But if the project to expand technological literacy has been justified as a means of achieving positive social change and new opportunity, to date it has failed to yield the significant positive social progress or productive changes that many people have come to hope for. Indeed, the American school system as a whole, and in the culture that this system reflects, computers continue to be distributed differentially along the related axes of race and socioeconomic status, and this distribution contributes to ongoing patterns of racism and to the continuation of poverty. It is a fact, for instance, that schools primarily serving students of color and poor students continue to have access to fewer computers than do schools primarily serving affluent students or White students. And it is a fact that schools primarily serving students of color and poor students continue to have less access to the Internet, to multimedia equipment, to CD ROM equipment, to local-area networks, and to videodisk technology than do schools primarily serving more affluent and White students. (p. 6)

Selfe tells a truth we're loath to admit when we have to think about systemic exclusions: they *still* exist, and they're *still* tied to race. Selfe's project as she outlines it in this part of her book, is significant not just because it begins to explore race and technology access. She also uses the book to examine the social and political implications that are involved in, but often hidden in conversations about technologies by looking at the Clinton administration's "Technology Literacy Challenge." What appeared to be a positive agenda by President Clinton, with its stated goals of "wiring every school in America," posed the danger, according to Selfe, of merely providing technology producers with the perfect market for their products: passive consumers dependent on those technologies, but unable to effect any meaningful change in the lives of those who have been marginalized.

Meaningful access to any technology involves political power and literacies: "in a formulation that literacy educators will feel most keenly, the project to expand technological literacy implicates literacy and illiteracy—in their officially defined forms—in the continued reproduction of poverty and racism. And it implicates teachers as well" (p. 7). That implication has become much more explicit in the years since Clinton's departure from office, because even those policies, with all their attendant dangers that Selfe points out, have been completely reversed. So not only did the Department of Commerce remove all of the "Falling Through the Net" reports from its Web sites immediately upon the arrival of the new administration and reduce the Divide to a case of class envy (as I'll detail later), but the Department of Education, stole and cruelly used the Children Defense Fund's "No Child Left Behind" slogan to sell policies that cut funding for schools that needed it most and identified thousands of schools serving poorer students and students of color as irrevocably "failing." But the subtitle of Selfe's book shows us just how much work remains for all of us in composition, computers and writing, and technical communication: "The Importance of Paying Attention." Both Selfe and Moran note that the field has not paid attention as of the turn of the century, imploring the field to finally catch up and take things digital and access to them seriously.

While race, and especially the connections between race and technology access serve as what Catherine Prendergrast would call the "absent presence in composition," conveniently contained within easy labels that prevent action on an anti-racist agenda (like tolerance, multiculturalism, and diversity, to take those in common parlance these days) (p. 36), the silence is not all-engulfing, as small numbers have begun to pay attention. There is important work being done with, and in response to, Selfe and Porter, even if that work still amounts to rumors and rumblings where strength and passion are needed.

Moran and Selfe, in an article published in *English Journal,* geared primarily to K-12 English and Language Arts teachers, "Teaching English Across the Technology/Wealth Gap," continue their call for teachers to place the Digital Divide at the center of research, teaching, and activism for writing specialists, and acknowledge the critical role race places in access to digital technologies. They also press on to identify important problems for those who take up the challenge. Selfe and Moran argue in the article that embracing technology uncritically can not only contribute to worsening the wealth gap and power differences between people of color and Whites, but also make teachers the best allies a global capitalist society could hope for:

> Advocates for technology often have an agenda that has nothing to do with our students' learning. If one is a politician or academic administrator in this decade, it is almost mandatory to call for technology in our schools, not because of any proven link between technology and learning—there really is no consistent evidence of such a link, especially in language arts and literacy studies—but because technology is seen as a quick and cheap fix for the perceived problems in our educational system. Anything associated with technology has a special glow these days. We note this in regard to the extraordinary bubble in technology stocks that his helping to drive the stock market .... In language arts and English classrooms, we need to recognize that we can no longer simply educate students to become technology users—and consumers on autopilot—without helping them learn how to understand technology issues from socially and politically informed perspectives ... [otherwise] we may, without realizing it, be contributing to the education of citizens who are habituated to technology but who have little critical awareness about, or understanding of, the complex relationships among humans and machines and the cultural contexts within which the two interact. (pp. 48, 52)

An investigation into the conditions of technology access and its relationship to education by the *Baltimore Sun* shows just how serious the problem is. Even when there is decent material access to computers, software, and internet connections, educators, students, and parents are still often hoodwinked, bamboozled, run amok, led astray, as Malcolm would put it. The fact that the grounds of officially sanctioned literacies are always changing, while students and schools in Black, Brown, and poor school districts are continually labeled as needing remediation as they navigate the seismic shifts in those grounds, in the high-stakes version of my electronic games parable in the prologue, leads to a collective desperation by parents, educators, and students themselves. Caught between the rock of political invective labeling them as failures and the hard place of magic pill solutions sold to them by technology companies, families and schools often make huge investments in technologies and still never escape the cycle.

The *Baltimore Sun* series "Poor Schools, Rich Targets" shows the devastation the convergence of forces Selfe calls us to attend: the Prince George's County, Maryland school district, in response to the pressure of being designated a school in need of improvement under reporting standards included in the No Child Left Behind Act, held trials for an Algebra software package it might purchase. According to the article, this package, named "Plato," costs on average about $60,000 per school. Huge investments like these are not abnormal. Camden, New Jersey's school district, with 19,000 students paid almost $11 million in site licenses to Compass Learning and a suite of educational video games from Lightspan. These programs are typically of the skill/drill/kill variety, and often only offer remediation. Compositionists can only wonder what happened to all of the intellectual struggle they waged over the last fifty years to move writing instruction beyond such a narrow focus, especially in schools and districts that have habitually reserved such instruction for poor students and students of color while children in Camden are stuck in labs working on grammar baseball games:

> One girl who was working on a punctuation drill was told to insert a comma in the appropriate place in the following sentence: "Robots will teach science in high school trade schools and kindergarten." The girl stared at the screen, which was illustrated with a picture of a robot, and then moved the mouse to place the comma between "school" and "trade." The screen lit up: She was right! Then it gave her another sentence to test the same skill: "The robots of 2019 will be able to see feel and have a better understanding of the world." Staring at the screen, the girl moved to add the comma to the correct spot again. (MacGillis, *Baltimore Sun*, 2004)

This is what Camden paid $11 million for, and the site licenses are only good for 3 years, after which time, of course, new versions will have been released costing even more. Administrators often feel that the expense is the only choice they have—a more justifiable use of funds than spending it on recruiting and paying signing bonuses to stronger teachers they often feel will not remain with the districts for any significant length of time.

Given these problems, Selfe and Moran articulate an agenda based on activism for equitable access across racial and economic lines, but for an access that is critically informed. This is the most important point for me, when we consider the ways Black students are continually dumbed down by skills only curricula, whether in the name of "raising standards" which usually results in remediation and those same exclusions based on Black students not being prepared enough for whatever the new dominant understanding of literacy in place at a given time. Technology issues must be as much the work of writing and communication teachers and scholars as writing and communication are, and we must find ways to build students' and teachers' digital literacies in the current environment of poorly funded schools and racialized education politics.

Toward the end of making the Digital Divide a central issue in curricula and pushing professional organizations to take public stands on technology policy issues, Selfe and Moran call or teachers to find ways to use any and all tools available in the project to expand access now. They suggest that given the expense of cutting edge technologies and the fact that there is always some new cutting edge software package or hardware tool being sold as the next great answer, the job of promoting digital literacies and writing abilities might often be best accomplished with lower-end tools (p. 52). Regardless of whether one pursues this strategy or not, the most important change in perspective for writing teachers is that they must make sure clearly articulated pedagogical goals drive all technology decisions so that purchases, training, and planning related to technology implementation remains relevant to the learning, social, political, and economic needs of those we hope to serve.

So a theoretical framework for initial action is in place in the field, if the voices offering it remain isolated ones. Jeffrey Grabill makes gestures toward filling an important void in computers and writing scholarship with his 1998 *Computers and Composition* article, "Utopic Visions, the Technopoor, and Public Access: Writing Technologies in a Community Literacy Program." This void is the need to engage some of the nascent policy/political/theoretical commitments in studies of how issues of access are played out in specific learning and teaching sites. Grabill's work is significant not only because it is concerned with access, but also because he argues that it is time for computers and writing specialists to take their work outside the composition classroom, to consider how "our understanding of how people write with computers will change because work in nonschool settings will alter inquiries and the knowledge produced" (p. 311). Grabill's understanding of access is built on Porter's definition (infrastructure, literacies, acceptance) and Porter's argument that access—who gets the resources, where they get access, what the environments will look like, and who will make decisions on all of these questions—is the key ethical issue that must drive all of our conversations about technologies and their relationship to written communication (p. 301). The Adult Basic Education program that Grabill studies serves primarily White women, but it clears valuable theoretical ground for composition and computers and writing faculty interested in community technology issues, and could link up in important ways with work on African American literacies outside the classroom as Elaine Richardson (2003) and Beverly Moss (2003) take it up in their respective books, *African American Literacies* and *A Community Text Arises.*

A few other pieces, mainly Susan Romano's (1993) "The Egalitarianism Narrative: Whose Story? Whose Yardstick?" and Kristine Blair's "Literacy, Dialogue, and Difference" advocate the use of computer-mediated communication technologies in the writing classroom, and attempt to disrupt simplistic assumptions that computers will magically create equal, democratic environments—moves that speak back in important ways to technology theorists like William Mitchell, whose *City of Bits* offers an architecture and urban planning model for the development of cyberspace with no consideration of the relationship between the racist histories of those professions and those who employ them and live and flee from the spaces they have designed and planned. Romano's and Blair's studies foreground valuable questions of Hispanic identity online, but are not overtly concerned with access.

One piece that does directly examine the connections between race and technology access as they affect African Americans is Elaine Richardson's "African American Women Instructors: In a Net." In her case study of three Black female writing instructors, Richardson works to show the political forces involved in their decisions to become more technologically literate, and to employ digital tools in their teaching. Richardson acknowledges the importance of technology access from the outset, but notes that women of color, because of the positions they often occupy as untenured or contingent faculty, do not have luxury of adopting critical positions on technological issues or "opting out" based on their critiques that other members of the academy, and particularly English departments, tend to have. This constraint, when combined with obstacles that hinder these three instructors from gaining meaningful access (increased teaching loads, graduate study, publishing requirements, family and other responsibilities) can make access, even when it is achieved, highly problematic. Richardson's study helps shed light on the complexities of access, showing that it certainly includes all of the elements of Porter's initial definition, but also extends beyond materials, skills, and community acceptance. The burden of access is not only the responsibility of those seeking it, but is a systemic burden as well.

Although many African American scholars and teachers are concerned about how new technologies might open up possibilities for resistance to racism and participation in American society on their own terms, the pressing need for these writers has often been to address the perception that African Americans just don't "do" science and technology. In "Computers on Campus: The Black-White Technology Gap," a *Journal of Blacks in Higher Education* article published in 1993, Terese Kreuzer quotes Richard Goldsby, an African American teaching Biology at Amherst College at the time: "They aren't likely to be hackers or computer nerds. Computers are not part of Black culture" (p. 93). A jarring comment, even if made 12 years ago, but one repeated many times in different ways in the scholarship on African Americans and science and technology. In another JBHE article, "Will Blacks in Higher Education Be Detoured off the Information Superhighway," Raymond Slater recognizes the difficult selling job in store for those who care about access:

> Black kids are not going to have the access to computers or online information services. We need to turn educators on to this problem. We are not sure how it's going to be done, but if it doesn't happen the technology gap between the races is going to widen. It's already wide, but it's going to get worse. (1999, p. 98)

Lois Powell (1990) begins an article in the *Journal of Negro Education* with the heading "The Black Scientist, a Rare Species." Some claim that there is no real problem with access to science and technology fields, but as recently as 2000, the National Action Council for Minorities in Engineering (NACME), a consortium of fellowship programs, faculty, foundation members, and National Science Foundation representatives working to solve this problem published *Access Denied: Race, Ethnicity, and the Scientific Enterprise,* examining the systemic factors limiting the participation of people of color in the sciences, engineering, and mathematics.

Knowledge of the systemic barriers that Black people have faced in the sciences and technology doesn't help much here. Goldsby's comment still echoes: "computers are not part of Black culture." However laced with the same kinds of self-defeating stereotypes about Black people that undergirded the Ebonics debate, Goldsby's concern points to the perception that the sciences and technology have never been central to African American struggles against racism and for equal participation in American society. Almost 25 years ago, Herman and Barbara Young (1977), in "Science and Black Studies," urged Black Studies programs to make study of the sciences a more central part of their curricula: "most Black studies programs have been concentrated in the fields of humanities, history, and the social sciences. They have neglected one area of special significance in achieving their objectives, which is the contributions of Black men to the fields of science" (p. 381). They argue that African Americans have a long history of achievement in the sciences, and that more focus on those achievements would help meet all of Black Studies' major disciplinary goals, including "provid[ing] alternative ideologies for social change" (p. 380).

A quarter of a century later, few African American Studies programs have answered the call. Abdul Alkalimat, African American Studies director at the University of Toledo is one of those who has, however, by explicitly linking technology to inquiry in African American Studies. In addition to creating eBlack, the online presence of Toledo's Africana Studies program, and in Alkalimat's words, the "virtualization of the Black experience," he uses the current moment's conversations about technology to keep reflections about what African Americans can and should do with technolog(ies) grounded in past struggle.

As he argues in an article published on cy.Rev (www.cyRev.com), "Technological Revolution and Prospects for African American Liberation in the 21st Century," the relationship between technological change and the economic, social, cultural, and political struggle of African American people is the missing link in the "history of African American history." Technological advances that improved the mechanization of the cotton and auto industries bring this relationship into relief for Alkalimat: "[c]otton and auto, as the leading sectors of the US economy—19th century agricultural and 20th century industrial production—helped to structure more than 150 years of African American labor. It has been this economic structure of how agriculture and industry have utilized African American labor that has set the stage for all of African American history." The perfection of the cotton gin provides the quintessential example of this relationship between technology innovation and African American struggle in Alkalimat's analysis, as it is the creation of the cotton gin that made cotton King and slavery even more profitable than it had been by the 1800s.

What makes this relationship important for the study of African American rhetoric is the fact that not only did these technological changes structure the conditions in which African American people lived, but influenced the environments in which they organized to resist those conditions. Just as the cotton gin mechanized the cleaning of cotton and created incredible demands for slave labor, Alkalimat notes, the mechanical cotton picker made sharecroppers obsolete, and was thus another important factor in what is commonly known as the "second great migration" of African Americans to the North (those migrating from the South after 1940) and led to conditions becoming favorable for the Civil Rights movement to emerge. He shows a similar pattern of high demand for African American labor being created, and then obliterated by technological change in the automobile industry, and points to the staggering unemployment statistics for African American men in the 1980s and 1990s in northern cities as evidence.

One need not agree with the breadth of Alkalimat's claim that "the most profound historical changes are linked to changes in technology" to take from his analysis an understanding that there are important connections between technological history in the United States and African American struggle. An important part of this connection is the relationship between communication technologies and rhetorical production. It is this relationship that I want to point to, as a way of extending Alkalimat's argument and making it relevant for African American rhetorical study.

Advances in communication technologies do not simply amount to minor changes in the medium. All technologies come packaged with a set of politics: if those technologies are not inherently political, the conditions in which they are created and in which they circulate into a society are political and influence their uses in that society (Winner, 1986), and those politics can profoundly change the spaces in which messages are created, received, and used. The potential of the changes in these spaces can be staggering when considered in light of the communicative possibilities they can either open up or shut down. One's ability to understand and operate within those changed spaces determines whether her or his linguistic dexterity is even relevant—as any faculty member who has been flamed (translation: dissed) on an e-mail listserv will be happy to explain.

Radio provides a better example: what it represents as a communication space and its potential as a medium for both entertainment and activism are not only dependent on the transmission of sound through amplified or frequency modulations or the broadcasting power that determines a radio station's reach. The politics of the Federal Communications Commission's decision to expand the numbers of stations individual companies can own in a particular market has turned a local medium into a national one and had effects ranging as wide a spectrum as wholesale changes in programming formats and the near elimination of the Disc Jockey as a relevant figure in the life of a community, beyond making paid advertising appearances for those businesses who can afford them. These effects produce both rhetorical problems and opportunities, as the Tom Joyner Morning Show demonstrates. Sophisticated rhetors and technology users must not only be aware of the politics that come packaged with technologies, but the responses they both enable and limit.

African Americans' ability to make the move from "ideology to information," and to take African American experience into new technological spaces, to digitize African American history, struggle, and celebration is not only important to survival for Alkalimat. It is also a move that offers possibilities that did not exist before: "[w]hile ideological struggle has persisted the information revolution has undercut the material conditions for ideological ignorance. The information revolution has increased our capacity to produce, store, distribute, and consume all texts—written, oral, and visual" (Alkalimat, 2001). eBlack, as a model of what is possible when African American Studies, traditions, and struggle, are digitized, is currently comprised of 5 programs or parts:

- Professional connection and discourse: eBlack edits and manages H Afro-Am, a listserv designed to connect faculty and graduate students in professional conversations in and about African American Studies. In the 3 years of its existence, it has attracted 1,000 members, and allows conversations that might never happen between scholars and students across wide geographic distances.
- Curriculum development through distance education: eBlack is developing a Pan-African studies program via the Internet, using distance education.
- Community Service: eBlack works to create websites for local churches in the Toledo area, and teach church members to create and maintain websites.
- Continuing Struggle, the African American Radical Congress: eBlack used the web to convene African American activists to "reinvigorate African American radicalism." Alkalimat notes that although many of the activists involved were skeptical about whether cyberspace could be an important site of struggle, organizing in this way not only allowed participants to document their work in spaces they could control, but that also prevented "factionalism and a hardening of ideological lines."
- Research Web site on Malcolm X: Alkalimat sees the web as an opportunity to combat the exclusions that are inherent in those institutions that have developed major archives of the papers of African Americans, like the Schomburg Center and many university libraries. The potential cyberspace provides for digitizing African American history can not only open up access to important figures like Malcolm X, but help to recover lost voices as well, including African American women, who Alkalimat notes, have been all but ignored by major archives.

What is important about Alkalimat's work with eBlack is not just that the histories and work of African American leaders, activists, and organization can be put online, but that cyberspace can transform that work because it is a space that, for all of the barriers to access that exist now, can allow more direct control of one's message than other media. As the Internet is currently constructed, a person who has a connection can view any site she wishes (that is not password protected, of course). Those with messages to share can target them directly to people they want to reach, unhindered by a television network's assumptions about African American people, or the limited range of AM radio or Cable television pubic access channels, for example. And it is far easier to own the means of production than it is a television station, radio station, newspaper, or magazine. So the possibilities for individual activists and rhetors are impressive, but they are just as bright for students and scholars of African American rhetoric.

Imagine, for example, if as a result of eBlack's work with churches in the Toledo area, or of any effort in any town, that twenty years from now, 2 or 5 or 10 churches in that area have Web sites that have archived all of their pastors' sermons (either in print, audio, video, or some combination of forms) over that 20 year period, with examples of church bulletins, directories, activities, and newsletters as well. One of the most studied forms in African American rhetoric, the sermon becomes much easier to study thoroughly, because many of the five challenges to offering rhetorical criticism of sermons that Lyndrey Niles identifies can be greatly minimized:

- Most African American sermons were not and are not prepared in manuscript form.
- Most African American sermons through the centuries were not and are not tape recorded during delivery.
- Some preachers are reluctant to release copies for criticism.
- Because most African American sermons are in dialogue form, manuscripts may not satisfactorily represent what actually took place in the church.
- Sermons in the African American tradition were not written to be read. Much of the real impact, therefore, is lost unless the critic knows how the words would have sounded, and can picture the delivery in his or her mind as he or she reads the manuscript.

As I noted earlier and Niles' problems show, African American rhetoric has always been multimedia, has always been about body and voice and image, even when they only set the stage for language. Even within a definition of African American rhetoric as being about the word, careful considerations of how current technologies can extend its study will provide a much richer body of work for rhetorical criticism and analysis.

This connection is not just about the politics of a particular writing, speaking, or designing space, nor is it just about the benefits that can exist for students and scholars of rhetoric. Technologies shape all five of the classical canons of Rhetoric, and greatly affect the rhetorical situation, however one might define the concept. Whether the technology involved is a soapbox, a megaphone, printing press, typewriter, telegraph, microphone, television, radio, or web site, the tools available affect the means available because of how they configure relationships between rhetors and their potential audiences.

Black intellectuals have always been concerned with improving African American participation in the sciences and technology, and to the degree that the humanities and social sciences are privileged, that emerges from the specific nature of the intellectual underpinnings of Western racism. Henry Louis Gates, Jr. and Nellie McKay (1997) look directly at this privileging of the humanities, offering some understanding of why literary production has had such a central role in African American struggle against racism for more than 200 years: "African American slaves, remarkably, sought to write themselves out of slavery by mastering the Anglo-American belletristic tradition" in ways "that both talked 'Black' and, through its unrelenting indictment of the institution of slavery, talked back" (pp. xxvii, xxviii). Literature—and literacies—hold this important place for several reasons, Gates and McKay contend. Black inferiority was a dominant theme in Enlightenment philosophy and science, as they show through extensive quotations from Immanuel Kant, David Hume, and Thomas Jefferson. Many people already know the story of Enlightenment justifications of racism, but the assumptions that Black people had "no ingenious manufacturers amongst them, no arts, no sciences" (p. xxx) was accompanied by a draconian body of public laws, making two forms of literacy punishable by law: the mastery of letters and the mastery of the drum" (p. xxix). Gates' introduction to *Race, Writing, and Difference* quotes a passage from Hume that makes these assumptions explicit:

> I am apt to suspect the negroes, and in general all the other species of men (for there are four or five different kinds) to be naturally inferior to the Whites. There never was a civilized nation of any other complexion than White, nor even any individual eminent either in action or speculation. No ingenious manufacturers amongst them, no arts, no sciences ... Such a uniform and constant difference could not happen, in so many countries and ages, if nature had not made an original distinction betwixt these breeds of men. Not to mention our colonies, there are negroe slaves dispersed all over Europe, of which none ever discovered any symptoms of ingenuity ... In Jamaica indeed they talk of one negroe as a man of parts and learning [Francis Williams, the Cambridge-educated poet who wrote verse in Latin]; but 'tis likely he is admired for very slender accomplishments, like a parrot, who speaks a few words plainly. (1986, p. 10)

Given this legal and intellectual history, it becomes easy to read African Americans commitment to literacy as a technological commitment and all of the means they have employed toward achieving literacy as technological mastery.

This connection between Black language and literary traditions and political struggle becomes even more explicit in Larry Neal's manifesto defining the Black Arts Movement. It is the Black Arts Movement that "is the aesthetic and spiritual sister of the Black Power concept" (p. 122). While the Black Arts Movement articulated goals for literary production that were very different from those of Phillis Wheatley or writers in the Harlem Renaissance, the larger purpose remained the same—to undermine Western racism based in pseudo science by using literature to talk back by talking Black, as

> writing, many philosophers argued in the Enlightenment, stood alone among the fine arts as the most salient repository of 'genius,' the visible sign of reason itself. In this subordinate role, however, writing, although secondary to reason, was nevertheless the medium of reason's expression. We know reason by its representations (Gates and McKay, 1997, p. xxx).

The fact that this literacy and literary activity was always engaged in a relationship with scientific and technological discourse cannot be overstated. V. P. Franklin and Bettye Collier-Thomas show this relationship in their reflection on Carter G. Woodson's (1969) work in founding the *Journal of Negro History*. They note that Woodson made it clear that this was to be "a quarterly scientific magazine" committed to publishing scholarly research and documents on the history and cultures of Africa and the peoples of African descent around the world. Woodson understood that publishing these articles and collecting these materials was the only way "that the Negro could escape the awful fate of becoming a negligible factor in the thought of the world." The activities pursued by the members of the Association for the Study of Negro Life and History (ASNLH) would "enable scientifically trained men and women to produce treatises based on the whole truth." V. P. Franklin used the life writings of African American literary artists and political leaders to demonstrate that "race vindication" was a major activity for Black intellectuals from the 19th century. African American preachers, professors, publishers, and other highly educated professionals put their intellect and training in service to the race to deconstruct the discursive structures erected in science, medicine, the law, and historical discourse to uphold the mental and cultural inferiority of African peoples (p. 1).

In other words, the chasm between literary and cultural production on one hand and scientific and technological pursuits on the other, is not a part of the Black intellectual tradition, at least for one of the greatest intellectuals of that tradition, Dr. Carter G. Woodson. But just as dangerous as assumptions that Black people are not technologically inclined, is the assumption that race is, and should be, irrelevant online.

The case I make throughout this book for an African American digital rhetoric probably seems contrary to some. It might even seem odd to attempt to argue that the entire African American rhetorical tradition can and should be read with a focus on the importance of technologies and African Americans' access to them. After all, the story of African American rhetoric to this point has been primarily understood as being about mastery of "the word," written and especially oral, throughout our history. Examples of this focus on speeches and essays abound, from foundational texts defining the scope of Black rhetorical study to articles analyzing specific rhetorical performances, regardless of the disciplinary homes (English Studies or Speech and Communication) of their authors. The titles of some of the major works in the field demonstrate this focus: Alice Moore Dunbar Nelson's (2000) *Masterpieces of Negro Eloquence,* Arthur Smith's (now Molefi Asante) *Voice of Black Revolution,* Philip Foner's (1975) anthology *Voice of Black America,* Foner and Branham's (1998) follow-up edition *Lift Every Voice,* Geneva Smitherman's (1986) trailblazing theoretical work *Talkin and Testifyin: The Language of Black America* and Gerald Early's (1992) collection on the African American essay *Speech and Power: The Afro American Essays and Its Cultural Contents from Polemics to Pulpit* are but a few examples. This is not to say that individual scholars in African American rhetoric have not discussed technology issues, but rather that traditions of Black mastery of spoken and written language and their use of it in collective struggle remains the frame that organizes the field.

Using such a frame works, obviously, for many reasons—not only because of its congruence with the events of that history from Frederick Douglass' ability to trick White playmates into teaching him to read when it was illegal for him to learn to Ida B. Wells' crusading uses of argument in "A Red Record" to condemn state-supported terror and lynchings of Black people, to Barack Obama's celebrated 2004 speech at the Democratic National Convention (a speech far more worthy of note for its poetic delivery than for any new Black progressive agenda, as it offered none)—but because it allows a relatively coherent view of that history. Because African American history has been so much about anti-racist struggle, because that struggle has been so consistently about making the argument for equal and meaningful participation in the nation, and because our greatest victories often emerge from the linguistic and argumentative brilliance of activists, poets, teachers, grassroots organizers, preachers, politicians, essayists, and musicians, it makes sense to organize that study around the spoken and written word.

It also allows for a somewhat stable narrative. One can link the power and commitment of Stevie Wonder's music to the development of Malcolm X's political voice and his ability to simultaneously take the best elements of the African American homiletic tradition and turn that tradition on its head. Carol Mosely Braun's fly signifyin on Jesse Helms in formal legislative debate (Crenshaw, 1997) can be connected to innumerable unnamed people across historical eras and social, political, and economic circumstances based on their use of argument to pursue a collective agenda on behalf of African American people, whether that agenda be one of resistance and revolution (Smith) or access to some notion of "the good life" (Golden & Rieke). Just as important, the word as the organizing trope of African American rhetorical study provides scholars and students with a ground on which to offer that narrative unshaken by the rapid changes that occur in communication media. It doesn't matter whether we talk about poetry that had to be authenticated by 18 of Boston's "most prominent citizens" because it could not be conceived that it was written by a African American person, as in the case of Phillis Wheatley, or the multimedia connectivity of Tom Joyner's attempts to get African American folks to "party with a purpose," the same analytical tools allow for conversations about all African American rhetorical production.

This promise of analytical stability across rhetorical situations and across communication media is a tempting one. So tempting, in fact, that to claim that technologies can be just as important a concept as the word is to take on a challenge, and might even seem to border on the heretical. Technology issues seem to be entirely unconnected to the making of arguments, and everyone knows how troubling African American relationships with technologies have been. But that's just the argument I want to make. Alongside our understanding of the centrality of language to African American rhetorical production, a parallel examination of how African American rhetors have used and manipulated communication technologies can foster both access and transformation.

As strenuously as I follow Abdul Alkalimat's lead, arguing that African American rhetoric has always been about technology issues, and that study in African American rhetoric should be opened up to examine more carefully those writing, designing, and performative acts that are acknowledged (but not appreciated as rhetorical in nature), I also argue that these other acts and texts need to be included *with*, not instead of oratory. I believe, with Philip Foner and Branham, editors of *Lift Every Voice*, that even now, with all of our technologies, oratory is still the genre of public engagement—the genre where the work any group of people commit to begins. As important as the Internet, books, magazines, journals, songs, dances, nightclubs, church services, and many other spaces are, it's still the sermon, the conference, the convention, the political speech, where the people make their commitments known. This is even more the case in African American communities, where Black people place an even greater emphasis on one's speaking ability, as those who are seen as African American public intellectuals themselves would have to attest. Had Cornel West never written any of the 20 books he has written, or written 20 more, his influence in African American communities would still depend to a large extent on his ability to "make it plain, reb," and make whatever knowledge he has translate at the podium. That said, work in African American rhetoric has to involve examining the technology issues involved in oratory at least as much as opening up Black rhetorical traditions to visual, electronic, and design issues.

Beth Kolko shows what is possible when scholars across this discursive divide do pay attention to each other, however. She argues that the history of racialized exclusions and their attendant assumptions of utter Black difference and irrelevance are programmed right into the interface of online environments like MOOs and MUDs—writing spaces similar to chats that accounted for much of the initial popularity of "cyberspace." Her essay, "Erasing @race: Going White in the (Inter)face," questions how race as a category has been elided in such media through various design choices, and it further investigates how the construction of "raceless" interfaces affects the communicative possibilities of virtual worlds (p. 214). This elision is a hallmark not only of the MOO and MUD environments with which Kolko is concerned—"the history of online communities demonstrates a dropping-out of marked race within cyberspace" (p. 214).

Kolko builds on the work of Richard and Cynthia Selfe (1994) with her examination of the interface as a raced yet e-raced space. This problem is important, in her words, because

> what this line of inquiry seems to represent is a growing awareness that technology interfaces carry the power to prescribe representative norms and patterns, constructing a self-replicating and exclusionary category of 'ideal' user, one that, in some very particular instances of cyberspace, is a definitively White user. (p. 218)

This is a very different conclusion about racelessness in online spaces than that reached by many theorists caught up in the initial technological rapture, who frequently assumed that cyberspace as raceless space would mean that race, gender, and economic class would no longer matter and would usher in new, equal, and thoroughly democratic worlds—both online and off.

Elaine Richardson and Samantha Blackmon are among very few African Americans in Rhetoric and Composition who address these issues. Their work, along with that of individuals like Kolko, Richard and Cynthia Selfe, James Porter, Jeff Grabill, and Johndan Johnson-Eilola, is important for several reasons. It takes technology access seriously as an intellectual and activist project for Rhetoric and Composition and Technical Communication instruction; it acknowledges directly the connections that exist between gender, class, and racism rather than attempting to reduce the Digital Divide to an outbreak of class envy, as has current FCC Chair (and son of U.S. Secretary of State, Colin Powell) Michael Powell; and it offers glimpses into an agenda for addressing the Digital Divide that goes beyond the mere existence of hardware and software to the quality of that hardware and software and Internet connections, technological literacies, and critical understandings of technology that are needed to gain meaningful access to any technology. Unfortunately, the fact that I can name the individuals who have attended to these concerns and review the work done to address them in the space of a decade has kept the Digital Divide firmly in place for Blackfolk and in the academy. In fact, we are guilty of exacerbating its persistence when, for a brief moment from 1995 to 2001, there existed some amount of political will to do something about it, even if the

Clinton administration's efforts to address the Digital Divide seemed based on a definition that amounts to the same access African American children have to public education or Black Floridians to the franchise: a vaguely articulated right to be in the same space *if* one navigates all the other barriers that prevent that presence. An examination of Clinton's policy initiatives on technology access and some of the shortsighted (sometimes well-intentioned, sometimes not) debates about whether a Divide even exists and how important it might be will begin to show one of the many ways technology access is a rhetorical problem and the need for our critical focus on it.

While one reason for the persistence of the Digital Divide is the reduction of African Americans to the level of people who have neither language, rhetoric, nor technology, another reason is the level of contest that exists around the term, especially when it is connected to race. Thus, the Digital Divide involves both contest and silence: debate over whether there is a Divide at all, waged by politicians, foundations, and business interests now; debate over whether race is a factor in whatever problems in technology access might exist; and concern that the use of a term like *Digital Divide* represents African Americans unfairly and does more to further the erasure of Black people by continuing to cast them as the utter outsider. These debates all carry assumptions about what constitutes access to computers, the Internet, or any digital technology that, even when guided by the best of intentions, threaten disaster if not addressed. This danger exists because those assumptions will guide legal, corporate, and educational policies that can trap Black people into roles as passive consumers of technologies rather than producers and partners, and worse, lead to continued electronic invisibility and economic, educational, and political injustice.

While African Americans have often at least implicitly understood technological issues as central to Black advancement, public recognition of race as a factor affecting access to digital technologies is a recent phenomenon, and largely due to the work of Larry Irving and many others on "Falling Through the Net," the series of U.S. Department of Commerce reports that introduce the term *Digital Divide* to the lexicon in 1995. "Falling Through the Net," in all four of its versions, acknowledged race as a factor just as important as education, income level, and geography in determining who is and is not able to obtain technology access. The first of the four reports, subtitled "A Survey of the Haves and Have Nots in Rural and Urban America," published in July 1995, examined access to the NII or National Information Infrastructure, by measuring the rates of telephone, computer, and modem "penetration" (i.e., the numbers of people who had telephones, computers, and computer modems in their homes). The major findings of the report were that education, age, and residence either in rural areas or central cities were important factors determining access to the Internet, but that race was just as crucial. Rural and city African Americans had the lowest rates of computer ownership, but the few who were online were significant in how they used the tools. Those African Americans were more likely than many other groups to use the Net to look for jobs, to take classes online, and access government reports. The policy recommendations of "Falling Through the Net" focused on connectivity: to develop more specific profiles of those who had telephones, computers and modems, and to provide assistance, primarily through an interim tactic of supporting community access providers like libraries and public schools. The fundamental assumption of the first report was the concept of universal access—that every household that wanted to connect to the National Information Infrastructure should, and ultimately, would be able to do so.

Three years later, the Department of Commerce and NTIA released a revision of the report, "Falling Through the Net II: New Data on the Digital Divide." The focus of this revision was primarily in providing a clearer look at the patterns that emerged from the first study, using 48,000 door-to-door surveys completed by the Census Bureau in 1997. In other words, the goals were to be able to better aggregate and disaggregate the data about who was connected and who was not, and to gauge the degree to which the patterns presented in the first report held or changed. The NTIA found that even with significant growth in computer ownership and Internet connectivity, the gap between numbers of African Americans who had access and the broader population *increased* between 1994 and 1997 (a difference of 21.2% in computer ownership rates in 1997 as opposed to a 16.4% difference in 1994). Further, as ubiquitous as telephone usage seemed to be for many during the 1980s and 1990s, there were still large gaps in telephone penetration across racial lines. Data collected for this report revealed that Whites were more than twice as likely to own a computer as Black people (40% vs. 19%) and almost three times more likely to be connected to the Net from home (21.2% vs. 7.7%). The policy recommendations in the second report remained the same; namely that more needed to be done to get people connected to the Internet, and that public access points would be an important means for providing that access. Again, the focus was on computer, telephone, and modem ownership as constituting access, and the report failed to engage the implications of Black, Latino/a and Native American people being disproportionately forced to share access in cramped public spaces while others worked in the comfort of home, with the leisure of learning skills and nuances of how the technologies worked. This assumption that mere physical proximity to a technology or ownership of the products is all there is to access is a troubling one that still dominates many studies on technology access.

The next of the reports, released in 1999 and subtitled "Defining the Digital Divide," concludes acknowledging the likely persistence of the Divide, and begins to offer data on computer and Internet usage in schools and libraries. The connections that some of us now take for granted between technologies, literacies, and education thanks to Cindy Selfe's book emerge here, as this version of the report warns the nation that it will be impossible to eliminate race as a problem in American society without addressing technology and information access. The report noted that economic survival would soon be tied to information access, and recommended policies and strategies aimed at filling the gaps that still existed in computer ownership and Internet connectivity, including lower prices, lease arrangements, and free computers, but took what amounts to a laissez-faire, free-markets-solve-most-problems pass given to those in the technology sector, arguing that increasing price competition between companies trying to gain market share would do much to minimize many of these gaps. This version of the report made references to issues of the relevance of technologies in peoples' lives and content (such as software and documentation), but these references brief and did more to skip over these problems than address them. The report called for outreach that would "let them know why they should care—how new technologies can open new opportunities for them and their children." To use an old missionary metaphor, salvation was only possible if the heathen saw the error in their ways and converted while government and business could continue to operate as usual.

The final version of the report, published in October 2000, sounds like a re-election campaign speech, proclaiming mission accomplished. It reported major progress among all groups, based on the same notion of connectivity (computer, telephone, and modem ownership) as the only important elements determining one's overall access, although it also considered computer use in public places as the third version did. The final report accounted more fully for how those with access used the Net, with information broken down into categories recognizing work, research, shopping, personal, and entertainment uses. Again, access is still defined as mere connectivity: not as literacies and community acceptance, as Porter argued for; not for critical awareness or change in the lives of individuals or communities, as African American rhetors have always argued for.

Other studies were undertaken to pursue some of the questions sparked by the "Falling Through the Net" series. Two of the more widely cited of these reports were Donna Hoffman and Thomas Novak's "Bridging the Digital Divide: the Impact of Race on Computer Access and Internet Use" and the Pew Internet and American Life Project's "African Americans and the Internet," both also released in 2000. While the Hoffman and Novak report acknowledges significant differences in access to the Internet by race, that access is also still defined by the access = connectivity model: questions focused on whether and how one used the Internet (in the last week, in the last 6 months),and whether that use took place at home or work, or at other locations. What is important about the Hoffman and Novak study is that, as the later NTIA reports do, it controls for education and income and shows that the Digital Divide is significantly connected to race. The conclusions from their work were that more African Americans were online than previously thought, and that African Americans need multiple points of access because home computer ownership rates are so different across racial lines.

The Pew report looked more at how African Americans used the Net than on questions of connectivity. Tracking a sample of 12,571 adults—2,087 of them African American—for 6 months, the study concluded that African Americans are more likely to have looked for job information, religious information, music, video, and audio clips, and more likely to have looked for places to live online than the general population. The Pew study, however, interpreted all of its results solely by comparison with the uses to which White Americans put computers and the Internet, repeating the erasure Beth Kolko argued is a fundamental characteristic of technology decision makers, making Whites' uses of these tools a technological default, rather than viewing the results through a lens focused on African Americans' needs, experiences, and histories with the technologies. The major result of the Pew study was that the Divide, in their view, was narrowing, and this conclusion has generated much conversation about whether or not people need to continue to view the Digital Divide as an "African American" problem.

Regardless of the problems definition of access operative in the initial "Falling Through the Net" studies, however, Larry Irving, its principal author, was very clear about trying to call the nation's attention to it as a problem of the legacy of racial exclusions in this country (no matter how muted the language might seem), calling it not only a major economic issue for Black people, but a Civil Rights issue as well. The reports on the Divide were used to support policies throughout President Clinton's administration that had universal access for all Americans as their stated goal. The potential these policies offered educators, even with all of their flaws in defining access as mere connectivity, was short-lived, however, because the concept of a Digital Divide that was deeply rooted in the stubbornness of racial exclusions came under attack from several quarters. Some politicians and public figures attacked the notion that such a phenomenon as a systematic Divide in technology access even existed, while others argued that its determining factor was class, and therefore, irrelevant, since, after all, class differences will always exist. Finally, even some Blackfolk committed to digital progress for African Americans grew weary of what seemed to be the same old, same old: more definitions of Black people as behind, and intentional silence about the technological innovation that has always been there, obscured by the story of a Divide.

In widely reported comments from his first press conference as the new chair of the Federal Communications Commission, Michael Powell maliciously reduced the Digital Divide to a toy fetish, announcing for all, from the beginning, that the Bush administration wasn't havin it, saying "I think there's a Mercedes Divide. I'd like to have one, but I can't afford one." He continued, "I don't mean to be completely flip about this ... but it shouldn't be used to justify the notion of essentially the socialization of the deployment of the infrastructure," a contention completely contradicting the "Falling Through the Net" reports' tepid strategies of wiring schools and libraries and its mild, responsibility-evading conclusions that the marketplace would eliminate any problems in access that remained. Of course, under Powell's leadership, the FCC would later mandate that televisions makers equip sets with high-definition technology rather than allowing free market forces that Republicans and Democrats alike often tout to allow consumers to decide whether they wanted such changes, a kind of "socialism" for Sony and Magnavox and other television makers, but such ironies did not matter when the FCC made the later decision. Intervention on behalf of technology producers to force innovation, force the obsolescence of millions of perfectly good television sets in the next decade was more important than intervention to help African Americans, American Indians, Latino/Hispanic peoples, and poor Whites in urban and rural areas achieve anything resembling "basic" access to the National Information Infrastructure. After referring to the "so-called Digital Divide," Powell added in the same press conference, "I think the term sometimes is dangerous in that it suggests that the minute there's a new and innovative technology in the market, there's a divide unless it is equitably distributed among every part of the society." And in what became the ultimate sign of disrespect, given the FCC's later policy decisions, he added "You know what? It's going to be the wealthier people who have the largess to go out and buy $4,000 high-definition TVS first. Does that mean there's an HDTV divide? No." Now, of course, no political leader, corporate executive, researcher, or teacher has ever argued that Mercedes Benz cars or HDTVs will be necessary for people to gain employment or participate in democracy, as politicians and legions of academics, community leaders and others have about computer-related communication technologies. However, this point was clearly irrelevant to Powell, who was content to dismiss all of these concerns and the nation's racist past and present as just the latest case of economic playa-hatin.

Not everyone challenging the concept of the Digital Divide or the racial element of the Divide are as polemical, political, or maliciously short-sighted as Chairman Powell was in his attempt to deflate the entire concept of the Information Superhighway (as the NII was called), to a high-end import car and a form of television the public still hasn't decided it wants to adopt yet, despite the FCC's activism on its behalf. Some African Americans in the technology industry have made moves to recenter conversations about the Divide away from race and toward class as the major factor. Barry Cooper and David Ellington are among those technology professionals. A Salon.com (Rawlinson, 1999) article quotes Cooper, the CEO of BlackVoices.com as saying that "people have legitimate concerns, but if [there] is a divide, it is economic." Ellington, founder and CEO of NetNoir.com concurs: "I don't feel there is much of a divide anymore. The Internet is becoming relevant in our lives as a result of e-mail and chat sites, and African Americans are going online in droves." In these cases, the same understanding of access as mere connectivity and usage, the same assumptions that drove the major early studies of the Divide were being used to dismantle the concept.

A collection of Black academics, known as AfroGeeks also have serious problems with the persistence of the Divide as a trope calling attention to the serious gaps in access that Black people continue to face. The group presents its motto as "from technophobia to technophilia" and is far more concerned with the ways discussions about the Digital Divide have "erased" the contributions of technology innovators. The group's home on the web, http://www.afrogeeks.com, hosted on the University of California at Santa Barbara server announces boldly:

> In recent years, African Americans, especially, have been portrayed as poster children for the digital divide discourse. Though rarely represented as full participants in the information technology revolution, Black people are among the earliest adopters and comprise some of the most ardent and innovative users of IT. It is too often widespread ignorance of African Diasporic people's long history of technology adoption that limits fair and fiscally sound IT investments, policies and opportunities for Black communities locally and globally. Such racially aligned politics of investment create a self-fulfilling prophesy or circular logic wherein the lack of equitable access to technology in Black communities produces a corresponding lack of technology literacy and competencies.
>
> Thus, necessary high-tech investments are not made in such underserved communities because many consider it fiscally irresponsible, which, in turn, perpetuates this vicious cycle. Despite such formidable odds, Black people continue to break out of this cycle of socially constructed technological determinism It is this way that African Disaporic people's many successes within new media and information technologies are too often overshadowed by the significant inequalities in technology access.

In other words, "you ignorant niggas is keeping me from getting mine, with all that noise about the access you ain't got." This seems like a harsh assessment of an argument that clearly speaks some truth, but the assessment is just for several reasons. The Digital Divide is not the equivalent of national "why Johnny can't read" campaigns that brand Black people illiterate in the service of continued American racism, but rather, was deployed by Black technology innovators in the Clinton administration (namely Irving and William Kennard, the FCC chair who preceded Michael Powell), community organizations, academics like Abdul Alkalimat, and activists in order to call attention to systemic inequalities—the same inequalities that the AfroGeeks acknowledge in their rant. Moreover, their embrace of technologies is unequivocal, lumping those with any critical consciousness about the ills they could cause or exacerbate as Luddites and technophobes who are part of the problem by announcing that the transition must be from phobia to philia. More worrisome than either of these issues, however, is the fact that they see their erasure as the *result of* the struggle for equitable access rather than part of the larger problem of the Divide, reflecting the same crabs-in-the-barrel mentality, the same scorn for Black people who have been systematically denied the educations and digital literacies they enjoy that were the worst excesses of turn of the 20th-century uplift ideology.

Although there are no absolute unities and many middle and upper-middle class Blackfolk will not have the same commitments as those who still struggle against systematic exclusion, middle class invisibility and the utter exclusion of working and poor Black people from the technologies, politics, and economies of mainstream American life still emerge from the same history of racism rather than problems we might have with each other. Rhetors like Marcus Garvey preached this incessantly, appealing for unity as we fight the issues that affect us rather than attacking each other over superficial differences. Unfortunately, these appeals for unity came out as bitter attacks on those Garvey saw as Black liberals imitating Whites, but the spirit of his argument beyond the rancor in his need to write himself into the discourse remain. In "An Exposé of the Caste System Among Negroes," Garvey (1923) writes "unfortunately, there is a disposition on the part of a certain element of our people in America, the West Indies, and Africa, to hold themselves up as the 'better class' or 'privileged' group" (p. 268). Garvey saw this developing caste as rooted in class, which at that point in time often meant skin complexion. Rather than skin politics, the issue here is digital politics, where the AfroGeeks charge that it is Black technology activists and their allies, rather than larger systems of racism that have denied them due recognition for their achievements.

I want to temper the charge just a bit at this point, because I do not believe the group intends to target Black activists as they do. It seems they misread the Digital Divide and the work toward equitable access it sparked as yet another in the long line of efforts I detailed earlier in the chapter, identifying Black students as Ebonics students needing remediation, and Black technology users and potential users as digital illiterates needing remediation. But to confuse Larry Irving and William Kennard for Rod Paige and Michael Powell is every bit as disingenuous—especially because the AfroGeeks are academics—as some of Garvey's well-known charges that W.E.B. DuBois worked against Black interests. By identifying "Digital Divide discourse" as the problem denying them their respect, the AfroGeeks risk the problem of seeming to try to write themselves into the larger digital discourse at the expense of young people and adults who can't eat or get decent jobs in an education system that has never yet answered the essential challenge of the Brown vs Board case and verdict from 50 years ago.

DEFINING ACCESS—AND A NEW VISION FOR AFRICAN AMERICAN RHETORIC

Fortunately, not all African American technology professionals view the Divide as Powell, Ellington, and Cooper do, nor do they all argue that we must stop talking about the Divide. And, in spite of the numerous debates that mark this territory, there is hope that those concerned with equal technology access for Black people can make that argument and demand that long traditions of Black technology innovation can be recognized. More sophisticated approaches to these debates can point us to higher ground, to more nuanced understandings of exactly what constitutes access, and a vision of what African American rhetoric might look like in this digital age. Let me begin this section with another version of the call I've issued:

> What is usually missing in our celebrations of African American history is a focus on how technological change contributes to the structural basis

> of African American history .... The entire sweep of African American history needs to be examined on the basis that technological change creates the main structural context for the grand historical narrative of enslavement and the subsequent freedom struggle. (Alkalimat, 2001a)

And one possible response, given the context of the Digital Divide we find ourselves in:

> The folks promoting this nonsense—I call them "tricknologists"—are the high tech equivalent of the three card monty dealers you see on street corners. You know the game: they get you to follow one card, and all the while the real action is somewhere else. Well, that's exactly what the New Age tricknologists are doing with the Digital Divide debate. The trick is simple: the first step is to narrow the definition of the Divide, by saying that computer ownership and Internet usage are how we measure minority participation in the new, high-tech economy. In fact, all these statistics prove is that minorities are closing the gap in becoming consumers of technology, not in being producers or equal partners.... Saying that the Digital Divide is closing because minorities have greater access to computers is like saying minorities have a stake in the automobile industry because they drive cars, or that they are Bill Gates because they own Microsoft Office 2000. (Taborn, 2001, p. 8)

You won't find too many engineers in African American rhetoric anthologies, but Tyrone Taborn, as publisher of *U.S. Black Engineer: Information Technology,* is obviously grounded in the tradition. His magazine is billed as "the African American community's technology magazine," continuing in the history of periodicals committed to advocacy for African American people. As is clear in the previous quote, taken from the February 2001 "Community Awareness" issue, Taborn understands technology as one of the major battlegrounds for African American struggle in this era of post-Civil Rights retrenchment. More than just an intriguing rhetorical performance in its own right, with its comparison of technology decision makers with everyday slicksters, Taborn's article shines a spotlight on the connections that have always existed between communication technologies, questions of access, and the African American rhetorical tradition. These connections can allow a new look at that tradition and how scholars might engage it in a new century, toward the combined ends of full participation in American society and the preservation and celebration of African American identity.

The history of African American rhetoric is, in many ways, one of individuals and collectives of writers, speakers, visual artists, and designers mastering, manipulating, and working around available technologies, even when access to them had been denied the masses of Black people. Through concentrating on this connection between communication technologies and rhetorical production, by looking forward to African American futures to look back through traditions of struggle, African American rhetorical scholars can not only offer far richer analyses of those speeches and writings considered to be within the tradition, but can also open those traditions up beyond just the word and show it has always been multimedia, using *all* the available means in resisting racism and pursuing justice and equal access on behalf of African American people.

Taborn helps us to understand that the AfroGeeks are partially right, but that Larry Irving and William Kennard are as well, as is Abdul Alkalimat in his work over the last decade. We must complicate access, and use our understanding of it to look forward, imagining Black futures, then back through African American rhetorical traditions and legacies of struggle, understanding histories to be created in the service of the futures we want.

What, exactly then, constitutes technology access, and how can it, whatever it is, sustain new visions of African American rhetoric, and rhetoric and composition more broadly? Before moving to that central point, I want to reiterate that this current context of technology access—of the Digital Divide and its connection to America's larger racial ravine—is an especially productive one in which to offer a new look forward to a new conception of African American rhetoric. This is because the Divide itself is a rhetorical problem at least as much as it is a technical or material one, and because technology issues have always functioned as a metaphor for imagining collective Black futures. As a field that has always had to address issues of access either directly or indirectly, African American rhetoric becomes central to both broader rhetorical and digital theory. And study at this important intersection of issues can continue the work of transforming the nation by eliminating that Divide, by addressing the real effects of technological exclusions, just as those rhetors we study always have. There are: the continual removal of Black labor from the workforce, inadequate health care, a legal system that operates on the presumption of Black guilt, a continual White flight into and out of urban centers, voting systems that conveniently do not work in major elections, and computer and information industries that operate comfortably under the assumption that Black people are non-technological while they spend billions of dollars to recruit, train, and develop temporary talent from overseas—when companies don't just send the jobs themselves overseas. All of this is at least possible—if technological issues make it onto our maps of the field, and if we end the silences that have prevented composition, computers and writing, and other elements of English Studies from taking Black intellectual traditions seriously.

A technological reading of African American rhetorical history offers at least three benefits beyond the obvious one of a more thorough appreciation of Black contributions to, challenges to, and even transformations of, the nation:

- A richer set of tools for analyzing those speeches and texts that have always been considered part of the tradition;
- A wider set of texts, images, sounds, and issues to address, and
- A chance to develop the arguments and policies that will end the Digital Divide and challenge the nation to accept responsibility for the exclusions pro-

> grammed into its technologies rather than continue to assume that technologies are neutral and that there is nothing systematic about the differences in American life that remain tied significantly to race.

Attempting a useful definition of meaningful access is a difficult task, but all of the partial answers that have emerged throughout this chapter show that it is possible: Porter's call for literacies and community acceptance in addition to the black boxes that dominate the Digital Divide reports; equal distribution of the black boxes themselves; the Selfes' and Charles Moran's call for critical awareness of the roles technologies play so that we avoid becoming passive consumers; even the AfroGeeks' insistence that we eliminate the centuries-old constructions of Black people as utterly illiterate and therefore unworthy of participation in the society and the technologies that govern it. All of these can contribute to an understanding of access that goes beyond the material and move us closer to the transformative ideals that unify Black rhetorical traditions.

One of the difficulties in defining access lies in the stubbornness of common understandings of technologies as merely the instruments people use to extend their power and comfort. Technologies also include the systems of knowledge we must acquire to use any particular tool and the networks of information, economic, and power relations that enable that tool's use. An example might help clarify this point. Law enforcement is a technological system for protecting the persons and property in a society, as well as the desired patterns of relations between them. Regardless of the availability of individual tools available to police in their work (guns, nightsticks, pepper spray, hands, feet, squad car computers, dashboard video cameras) and the wide range of force those tools represent, young Black and Latino men (and increasingly, women) are killed, injured, arrested, charged, and convicted at higher rates than other groups of people in this country.

Observers from different academic disciplines would offer very different explanations for why these patterns endure, why as a nation we were not horrified enough by the cases of Amadou Diallo and Abner Louima to demand fundamental change in policing, although we were (I use *we* sarcastically here) angry enough to press for changes to the judicial system in the aftermath of the Simpson verdict. I have no desire to take up those disagreements here. My point is simply this: any decision about how police use the tools and force they have is largely a result of what they have been taught about when and how to use them and the mandates police forces are given that construct crime and criminals in particular ways. Because of these literacies and processes and encoded forms of knowledge, one will almost always hear a defense of officers being within the law or policy of using the "maximum allowable force" in answering a brutality complaint rather than being expected to use the minimum force necessary to subdue a suspect.

The Clinton administration efforts I've cited throughout this chapter to "wire every school in America" in the mid to late 1990s offer another example. Many technologists and educators argued passionately that computers and the Internet connections in schools would be the great equalizer in American education. Ensuring that all students had networked computers in their schools would not only avail them of information that had never been distributed equally in their land-based schools, but would make their teachers better and motivate students to learn. For all of the passion that went into creating programs and writing grants to install hardware, software, and connections in schools, no mechanisms were in place to ask how the new computers would be used. Few resources existed for school systems to teach either students or faculty how to even use these tools, much less integrate them into anti-racist democratic teaching agendas. Few considered the difference that anywhere from a 5 to a 25 year head start would make for teachers, students, and parents who had already been indoctrinated into digital culture, and more importantly, had been made comfortable enough with computers and the Net to take advantage and ownership of the technologies. And when people did begin to consider these issues in the last few years, almost no African Americans were present where the decisions were being made, either in technology companies, policy think tanks, university faculties, consulting companies, vendors, IST or technical writing faculties, or the policies, documentation, content, interfaces and environments, or training efforts any of them created.

The problem with the Digital Divide as a concept for addressing systematic differences in access to digital technologies is that it came to signify mere material access to computers and the Internet, and failed to hold anyone responsible for creating even the narrow material conditions it prescribed. Beyond the tools themselves, meaningful access requires users, individually and collectively, to be able to use, critique, resist, design, and change technologies in ways that are relevant to their lives and needs, rather than those of the corporations that hope to sell them. Let me sketch what I believe is a more effective matrix for understanding technology access and then conclude with some reflections on what this changed understanding might mean for writing instruction—the work of composition, computers and writing, technical communication, and African American rhetoric.

Of course one has to own, or be near places that will allow him or her to use computers, software, Internet connections, and other communication technologies when needed. The "Falling Through the Net" reports, along with many other studies brought this need to our attention. I call this *material access,* and meaningful access begins with equality in the material conditions that drive technology use or nonuse. To really play out the implications of this statement would mean that there can be no real digital equality without fundamental transformations of the economic relations in our nation. But even that "revolution" alone would not be enough.

For material access to have any effect on people's lives or on their participation in the society, they must also have the knowledge and skills necessary to use those tools effectively, or what I'll label as *functional access*. I use this term because, just like functional literacy, it is insufficient for economic or political power or for many kinds of participation in the nation's social or cultural structures. Porter's call for us to understand access as involving the skills and knowledge necessary to benefit from technologies is partially a call for this kind of access.

Porter also notes that people must actually embrace the technologies involved, that there must be a level of community awareness and acceptance in order for those technologies to mean anything. Beyond the tools themselves and the knowledge and skills necessary for their effective use, people must actually use them; they must have *experiential access,* or an access that makes the tools a relevant part of their lives. In addition to discerning relevance in the technologies, people must have some involvement in the spaces where technologies are created, designed, planned and where policies and regulations are written. They must be present in the processes by which technologies come to mean what they mean for us. This notion of experiential access is a kind of fusion of Porter's understanding of access and Taborn's insistence that access also includes power in the creation and shaping of our technologies.

Richard and Cynthia Selfe's work, along with that of Charles Moran show us that not even these layers of access, alone, are enough. Members of a particular community must also develop understandings of the benefits and problems of any technology well enough to be able to critique, resist, and avoid them when necessary as well as using them when necessary. School districts must know when the wonderful sales pitches of computer companies—and maybe even their donations—won't work for their budgets, curricula, and strategic plans. I term this ability *critical access.* Let me offer what might seem to be a simplistic non-technological example here. Just as we make decisions about what foods are healthy for us or not healthy for us, or which ones fit our bodies' needs at a particular time when at a supermarket or restaurant, any group of people must know how to be intelligent users and producers of technology if access is to mean more than mere ownership of or proximity to random bits of plastic and metal.

Supermarket savings cards provide a more relevant—and seemingly ubiquitous—technological example. These cards, offered by large supermarket chains, are used to track consumer purchases. These chains then use, sell, and trade this information with many other corporations in order to create demographic profiles of communities and make decisions on whether to expand or withhold products and services to them, just as credit card companies and other corporations do. Oscar Gandy describes the potential for racism in the ways corporations gather and use this information in his article "It's Discrimination, Stupid!" Given the obvious value and use of this information and the way supermarkets virtually force consumers to sign up for these cards with pricing strategies that make any product savings dependent on one's ownership of a savings card, and given the pricing disparities that often exist in grocery stores that major chains operate in inner cities, African Americans and many other groups of people face important decisions about whether or not they should own or use these cards, and if they do, under what terms. Just as Operation Breadbasket organized consumer education and boycotts in the 1970s to encourage Black consumers to withhold their business from companies that did not hire Black people, it is past time for African Americans to engage in targeted and mass protests and boycotts of supermarkets, science and technology departments in universities, and corporations in the technology sector that fail to recruit, hire, or promote African Americans and members of other underrepresented groups while they spend billions of dollars on myriad programs (including the H1B visa program) to recruit technical talent from overseas.

The understanding of access I take here echoes the language of literacies that has become common even in non-academic circles, and is informed by Stuart Selber's and Johndan Johnson-Eilola's (2001) framework of "thinking, doing, and teaching" that they argue can provide the multidimensional basis that instruction in technical communication needs to better prepare its students for the complexities involved in the technologies with which they will live and work (p. 4). It is also one that locates technology as a site and means for African Americans' continued efforts to "carve out free spaces in oppressive locations such as the classroom, the streets, the airwaves" (p. 16), as Elaine Richardson documents as the purpose of other Black literate practices. In other words, technology is both one of those sites of struggle and a possible means of liberation, something we can not only survive but transform in our own interests.

This taxonomy of access that I have begun to piece together—to quilt—from the patches of others' definitions and debates is best thought of rhetorically as taking place along the related axes of critique, use, and design. The larger project of making access meaningful for African Americans and other marginalized groups must combine participation, resistance, and the re-creation, the reimagining of technological systems, artifacts, processes, and the economic, power, and social relations that are embedded within them. Table 2.1 begins to establish the relations between the items on my list and the traditions of critique, use, and design that often govern explorations at the intersection of rhetoric and technology. The questions I include within it represent only the merest of beginnings, but include issues I address later in the book as well as concerns taken up by others like Tyrone Taborn, Cynthia Selfe, James Porter, and others.

It bears repeating at this point that computers, the Internet, and other technologies included in the Digital Divide are only a part of the larger racial ravine—the complex of technological divides that have always operated in American history, and have always been used to maintain exclusions connected to race. The ways racial exclusions are built into our technologies and into our society operate to continue and even ensure those exclusions while allowing individual members of that society to avoid responsibility for that exclusion. This helps to explain why the computer industry—the only major American industry to truly emerge in the aftermath of the Civil Rights and Black Power movements—remains the most segregated in the American workforce. At the same time individual companies continue to congratulate themselves for having inclusive advertising campaigns or for funding individual "minority" science and engineering programs or offering occasional grants and "legacy" (translation, 2nd, 3rd generation or older) machines and stripped down education versions of software programs in order to foster brand loyalty.

TABLE 2.1

Critique	*Use*	*Design*
What are the patterns of computer and technology ownership by race? How are those patterns over-determined by present patterns of racial exclusion? Where do African Americans use computers, and how do the policies and assumptions about education at work in those spaces affect the uses that are available to them? (material access, critical access, functional access, experiential access) What economic and policy factors influence those patterns? (material access) What tools do activists, scholars, teachers, users have available to interrupt and intervene in closed technological systems? (critical access)	How does the distribution of a technology affect how people use it? (material access) What cultural retentions do African Americans and other people of color bring to the technologies they use? What relationships exist between these retentions and the acquisition of digital literacies? (functional access) How can technical communication address race and culture more effectively in areas like the writing and design of documentation and policy? (functional access, experiential access) How can educators avoid being swindled by technology companies eager to manipulate their desperation? How can they make better spending decisions with the limited resources they have? What pedagogical aims direct technology spending choices? (functional access, critical access) How does one decide when to promote low-tech solutions to specific users and learners needs? (critical access, material access)	How can the aesthetics and functions of technological artifacts and interfaces be made more culturally relevant? How can the cultural relevance of design questions like this affect the degree to which users embrace technologies and take ownership of them? (experiential access) What specific visual and design traditions can African Americans bring to technology design? How can African Americans counter the design processes and practices of technology firms that have rigorously excluded them? (critical access)

Ownership of computers and other digital technologies, knowledge of their uses, participation in their processes, and even critique of their failings are not enough either, however. The way of looking at race and technologies, and of a meaningful technology access, goes further; to transformation. This kind of look at race and technology posits that our nation is a construct, or system, maybe even a technological system, and the Constitution, federal, state, and local laws, as part of the code that runs it. Social spaces like schools, cities, the workplace, and the court system are all interfaces where people use that system. African American struggle as reflected its rhetorical traditions, was always an attempt to *both* change the interfaces of that system *and* fundamentally change the codes that determine how the system works. My exploration of technology issues in African American rhetoric, then, is intended to document the ways Black people have hacked or jacked access to *and* transformed the technologies of American life to serve the needs of Black people and all its citizens. My goal here is not merely to encourage long-denied appreciation of these traditions of struggle, but to enable new (as well as ever present but unrecognized) strategies and tactics for pursuing African American freedom in this recalcitrant racial moment.

African American rhetoric has been about all of this and more; Black people have continually worked for a *transformative access* to all of the technologies that make up American life. By transformative access, I mean that African Americans have always argued for a genuine inclusion in technologies and the networks of power that help determine what they become, but never merely for the sake of inclusion. African American rhetorical practices call attention to the ways that the interfaces of American life, be they public facilities, education, employment, transportation, the legal system, or computer technologies, have always been bound up in contests over language, and have always been rhetorical—about the use of persuasion, in these cases, toward demonstrably tangible ends.

One might argue, that at best, the United States has adopted what we can call a "user-friendly" racism in the aftermath of Black freedom struggles of the middle and late 20th century: a language of tolerance and facile efforts to make individual interfaces more accessible (such as public facilities and education) in a system that remains fundamentally unchanged, and in constant contest over even those minimal efforts. Just as Civil Rights struggle was about far more than the right to sit next to White people at lunch counters or in schools, technological struggle in this century must be about much more than merely owning or being near computers or any other technological tool. The question is not about how many of our schools are wired, but rather, as Martin King put it in one of his final speeches "Remaining Awake Through a Great Revolution," the problem of how "through our technological and scientific genius, we have made of this world a neighborhood and yet, we have not had the ethical commitment to make of it a brotherhood ... [we live with the] unhappy truth that racism is a way of life ... spoken and unspoken, acknowledged and denied, subtle and not so subtle" (1991, p. 270). King's point in this speech, which I take up in more detail in the next chapter, is that the racism in our society infects our technologies just as it infects our media, our schools, our politics, our economy, and that the technologies themselves must be transformed. What does this transformative ideal look like? What set of attitudes, commitments, and goals might enable one to pursue this kind of transformation?

3

Martin, Malcolm, and a Black Digital Ethos

There are perhaps no better examples of the important ways the African American rhetorical tradition is shaped by technological issues than the two figures in that tradition who attract the most scholarly and public attention. Thousands of books and articles have been written for scholars, students, and the general public that speak to the rhetorical brilliance of Malcolm X and Martin Luther King, Jr. One might go so far as to say that their combined persuasive brilliance accounted, in part at least, for the rise of African American rhetoric as a distinct field of study. Their speeches and writings have been collected, anthologized, and analyzed from a seemingly endless range of perspectives.

That discursive brilliance deserves every bit of the attention it has received and more. There are and were many in Black communities all over space and time who had King's or Malcolm's oratorical gifts. However, few if any, of those books and articles accounts for King's or Malcolm's savvy with media like television and radio, or the broader understandings of technological issues as crucial to African American struggle. As valuable as the innumerable speeches, sermons, letters, editorials, essays, and books that they wrote were, one could argue that the most important rhetorical achievements in both King's and Malcolm's careers occurred on television. What catapulted Malcolm X into the national imagination and transformed the Civil Rights movement from a set of disturbances led by a "rabble rouser" to a coherent national movement led by one who would become a Nobel Prize winner and national hero depended thoroughly on Malcolm X's and King's grasp of how television worked as a rhetorical tool. What continues to make their rhetorical brilliance relevant for us in a new century is their incisive analysis of the roles technological issues played and still play in maintaining the exclusions we continue to face. This chapter looks at the leaders' ability to both manipulate individual technologies as a part of their rhetorical mastery and bring a critical perspective to broader technological issues. This combination of mastery of individual technological tools and more general theoretical awareness comes together in what I argue needs to become a Black digital ethos—a set of attitudes, knowledges, expectations, and commitments that we need to develop and teach and bring to our engagement with things technological. Martin and Malcolm embody this ethos and serve as models for ways to deal with the complex relationships that are involved in the attempt to develop an informed, critical perspective to the project of transforming a nation in this digital age.

In the case of Dr. King and television, King knew that because of national television news, he had a vehicle for demonstrating, almost instantly, the extreme nature of White racism and the moral justice of Civil Rights protesters and their cause. This vehicle stood in stark contrast to the monthly or weekly publication cycle of many newspapers and magazines, though he obviously depended on print media as well—particularly the pools of print journalists who followed the cameras. More striking than the relative speed with which these messages could be carried was the mere fact of video footage: instead of still photographs, or the verbal and written descriptions of writers and television anchors, Americans could see the action happening. For all of King's eloquence with the written and spoken word, it was a visual rhetoric of innocent protesters being beaten and hosed, of callous politicians putting extreme words with those hoses and beatings, that made the appeals successful. That footage helped convince a disbelieving nation that racism was really that vicious, that brutal. In spite of all of the work that has been done on the intellectual and philosophical underpinnings of King's tactics during the Civil Rights movement, studies exploring the rhetorical underpinnings of his radical, aggressive nonviolence as they related to what was still an emerging technology at the time are nearly nonexistent.

The same argument could be made for Malcolm X. Although he was verbally brilliant, the Nation of Islam has always had verbally brilliant mosque leaders and spokesmen. In spite of the talents of many of its mosque leaders, however, the Nation had been dismissed as either an irrelevant cult or an extremist group for the most part, even early in Malcolm's career. One moment, carried on television, is at least as important as many of his famous speeches to how we view him as a rhetor, and possibly changed his role in the movement and the Black imagination. On July 13–17, 1959, the television special "The Hate that Hate Produced" aired nationwide, bringing Malcolm X and the Nation of Islam to the public's attention.

Although a person who is interviewed on a television show does get a chance to voice his or her views, that person often has very little control over that voice. In the case of Mike Wallace's special on the Black Muslims (as the Nation of Islam was called at the time) and Malcolm X, Malcolm was presented on a television show that writers, producers, directors and television executives planned and approved. Contemporary relevance is obviously a factor in why networks pursue news specials, and this relevance is often calculated by the number of potential viewers such a special might attract.

Once a concept is approved, the special is planned, footage is collected, research done, the episodes of the series mapped out, the reporters' narrative written, and interviews are performed. All of this takes place before the public sees the individual invited to appear on it, even when the special is live. Decisions are made about what to include and exclude, even from individual interviews, how those segments will be arranged, and how the reporter will comment on them, both before and after. The interviewer knows long before the subject what questions will be asked, in what order, and how much time the subject will be given to answer—regardless of whether or not they are generous enough to brief the guest on those questions in advance. After all of that, the interviews, the speeches, or whatever else might be presented on an individual or organization is edited. This editing is done with consideration for time constraints, commercial slots, the interviewer him or herself (after all, any news show is more about the anchor or "talent" than any individual guest), and the biases of those involved in the process.

With all that in mind, consider the television landscape for African Americans in 1959, if in 2001 organizations like the NAACP still have to threaten boycotts of networks over hiring practices and issues of representation. Even now, it is nearly impossible to find print, broadcast, or even Internet coverage of African American issues in which those issues are taken seriously. Amy Alexander deals with this ongoing problem in an Africana.com article in which she interviews three Black women journalists, Jill Nelson, Julianne Malveaux, and Tananarive Due. All four journalists comment on the ways issues facing African Americans were completely dismissed during the 2004 presidential election, even as both parties spouted claims that the African American vote would be crucial to the election's outcome. This dismissal was so complete that even when Gwen Ifill asked vice-presidential a question about Black women and AIDS, neither candidate had even the beginnings of a response, although each candidate for president and vice president had thoroughly rehearsed (if not substantive) answers to questions on any other subject. This dismissal of the concerns of Black people in broader policy discussions inside and outside of political campaigns is connected to the sickening absence of Black women and men in mainstream news media organizations, print, television, and online. Even the potential that the Internet would seem to offer through Web sites and blogs is disheartening at best. Julianne Malveaux describes it best, saying

> this campaign has been nothing but pack journalism. There has not been a lot of thoughtful coverage, and there's been way too much hype. The addition of bloggers to this mix ... [and the high level of coverage by the New York Times and other outlets] has been ridiculous ... these bloggers push it as if they are adding new dimensions to the media. They are simply younger mirror images of the White boys of the bus. Of the 35 bloggers given convention space at the Democratic convention only one was African American. (http://www.africana.com, Nov. 1, 2004)

Television, 1959: what role did African Americans play in the overall business of this communication space? What writers, engineers, videographers, reporters, anchors, directors, or producers worked on television? Where could an African American leader go to present a message on television where African Americans played at least some part in deciding how that message would be shaped and received? Given that broader rhetorical situation, how does one answer questions when the interviewer controls the questions and the conventions of turn taking, and an entire staff overtly shapes the context before the interview happens? How does one plan a message and an argument when the program's staff has the power to edit as it sees fit, without consulting the person interviewed? Even on live television, what strategies does one employ when the show is paced in such a way that the interviewer gets the first, last, and arbiting words? In such a situation, how did Malcolm manage the interviews in spite of this tremendous difference in access and power? How did he respond to what he knew was a stacked rhetorical deck on a four day special whose title took for granted the widespread assumptions that he and his group were hatemongers? How can these considerations help us to appreciate even more Malcolm's cool under pressure on the program, his incredible quick wittedness, and argumentative focus in his spoken answers to the questions asked him?

Clearly, "The Hate that Hate Produced" cannot be reduced to merely an example of White racism run amok in television news, even as it puts into high relief the stakes involved when African Americans and other excluded racial groups hold no control over media outlets. Too many other issues prevent such a simplistic reduction: many African Americans were just as perplexed and incensed about the Nation of Islam as a group, Malcolm X as its spokesman, and Black nationalism as an ideology as White America was at the time. Louis Lomax, the Yale PhD who conducted the interviews for the series, was very critical of Black nationalist ideology as well. Lomax was one of a handful of integrationists Malcolm X respected, however, as he was willing to engage Malcolm on the limits and possibilities of integration and Black nationalism and the Nation of Islam's particular version of it, accepting Malcolm X's challenges to public debate. Lomax often wrote thoughtfully on the Nation and Black nationalism in books like his 1963 examination of the Black Muslim movement *When the Word is Given: A Report on Elijah Muhammad, Malcolm X and the Black Muslim World*. Lomax was also present in Cleveland at Cory United Methodist Church when Malcolm delivered his classic speech "The Ballot or the Bullet," discussed later in this chapter. Even as Lomax's interviews followed the general slant of the program, his careful questioning is completely drowned out by Wallace's constant editorializing. Although Lomax is clearly the expert on the Black Muslims and is the one who did the research and reporting for this documentary, Wallace is provides the only interpretive framing for the interviews that follow. Lomax is allowed a short interview with Wallace, but the realities of American television of the time were such that Lomax would have no opportunity to comment on his findings himself, no chance to provide any of the perspective that would have helped balance or temper the anger that the Black nationalists in the program incited, no moment in which to call viewers' attention to comments by leaders like Adam Clayton Powell meant to bring theological and ideological complexity to the case made by the Black Muslims, even as Powell disavowed the incendiary language in which their messages were wrapped. The conditions of access and the conditions of race were such that no other story about the Nation of Islam, Elijah Muhammad, or Malcolm X *could* have been told on network television, even during the era hailed as the height of Edward R. Murrow-esque commitments to objectivity and neutrality.

We're used to subtle and sophisticated slant from media organizations today. Mike Wallace and those who helped to plan, research, and produce this documentary were under no such requirement. The widely held assumption that Malcolm X, the Nation of Islam and all Black nationalist groups of the period were "Black supremacists" and hated all White people was the central idea of this television special, and all of the interviews and background information presented within it were done so ostensibly to prove that this assumption was indeed true. Wallace repeated assertions that the Nation of Islam was dominated by a hatemongering ideology that was dangerous to the nation literally dozens of times in his commentaries introducing segments with absolutely no attempt to balance this perspective. This lack of balance in the presentation of what Wallace calls "this disturbing story" occurs in the special with complete disregard for the fact that Malcolm and others insist that they do not hate all White people. Groups like the Nation of Islam are contrasted with "sober-minded Negroes" and said to be preaching a gospel of hate that would "set off a federal investigation if it were preached by southern Whites." The major evidence, if it can be called such, is based on a skit or morality play often presented at mosques and public events called "The Trial," using the courtroom metaphor to indict White America for what it saw as a deeply rooted hypocrisy. "The Trial" is similar to other nationalist arguments made by many rhetors, most notably David Walker's *Appeal,* where the focus is not only on the nature of Black oppression and the ways African Americans have been summarily labeled as immoral or savage, either overtly or covertly in justifying that oppression, but on the extreme nature of both the oppression and the hypocrisy involved in the racist attitudes at its root. Just as David Walker repeats over and over again that slavery and oppression against Africans in America

> are the most degraded, wretched, and abject set of beings that ever lived since the world began; and I pray God that none like us ever may live again until time shall be no more. They tell us of the Israelites in Egypt, the Helots in Sparta, and of the Roman slaves, which last were made up from almost every nation under heaven, whose sufferings under those ancient and heathen nations, were, in comparison with ours, under this enlightened and Christian nation, no more than a cipher—or, in other words, those heathen nations of antiquity, had but little more among them than the name and form of slavery; while wretchedness and endless miseries

> were reserved, apparently in a phial, to be poured out on our fathers, ourselves, and our children by *Christian* Americans! (p. 23)

The commentary in "The Trial" is very similar:

> I charge the White man with being the greatest liar on earth! I charge the White man with being the greatest drunkard on earth! I charge the White man with being the greatest swine-eater on earth! (yet the Bible forbids this) I charge the White man with being the greatest gambler on earth! I charge the White man with being the greatest murder on earth! So therefore, ladies and gentlemen of the jury, I ask you, bring back a verdict of guilty as charged! ("The Hate that Hate Produced")

Other supposed faults of the Nation of Islam "the most powerful of the Black supremacist groups ... let no one underestimate their influence" lie in their attempts to develop Black economic power, educational institutions, and cultural resources. Much of the rest of the documentary is a trial similar to the one staged by the Nation. Witnesses are brought out and asked to account for their comments about race, racism, and the Nation of Islam's views of White people. Elijah Muhammad and Malcolm X are interviewed after commentary about court charges each faced and jail sentences each served, casting its leader and spokesperson, and therefore the organization itself, as criminal. After Muhammad and Malcolm X are interviewed, some of the "saner Negroes" are called in and, in a sense, compelled to testify against what is presented as the obvious racism and hatred in the Nation's ideology and teachings. Guests include James Forman, Adam Clayton Powell, and others called in mostly to condemn the rising popularity of various forms of Black nationalism and Elijah Muhammad and Malcolm X in particular.

One of the more interesting elements of this documentary is the reporting and interviewing work of Louis Lomax, an African American intellectual who also worked as a reporter for various Black newspapers before his work on Wallace's special. Although Lomax worked diligently to question his subjects fairly and to leave them space to offer their own answers in their own terms, his questioning is clearly shaped by his charge to document the "hatred" espoused by the group, reinforced over and over again by the footage selected and Mike Wallace's incessant commentary that provided all of the framing for the interviews. In spite of Lomax's sensitive questioning, and his knowledge of American and African American history, Malcolm X and anyone else on the program was faced with a stacked rhetorical deck, or a situation where he was forced to attempt to present his views when the evidence was already in and the trial was already decided, to continue with the metaphor. When individual interview subjects like Adam Clayton Powell or Dr. (insert name) offered thoughtful, incisive analysis about the conditions of race involved in the emergence of Black nationalist groups or offered subtle distinctions between the sensationalism that often accompanies rage and the hate at the root of White racism, the implications of these comments are ignored.

What might this series of interviews, this investigation of an organization and the emergence of nationalist ideologies, have been if Lomax were the series' producer, had access to African American camera crews and research staff, and the freedom to present his results to an African American audience unfettered by the restraints of overbearing White public opinion that was already formed and unwilling to be persuaded? As Amy Alexander and the journalists she interviewed suggests, we cannot know the answer to this question even today. So the question then becomes how individual rhetors deal with that stacked deck when it is the only deck available if one is to be heard at all.

What is fascinating about Malcolm's use of the medium might best be presented by contrasting his interview with that of Elijah Muhammad. Muhammad is reserved and often cryptic during his interview with Lomax, his answers matter of fact, and often simply yes and no answers. He exhibits a direct style, answering questions forthrightly, but shows no desire to engage in dialogue with those who disagree, or to explain or make a case for the Nation of Islam or its beliefs.

Malcolm X's interview differs sharply in that while Malcolm is also calm and sometimes even reserved in his demeanor, he also explains diligently elements of the Nation of Islam's ideology and theology, and relishes the chance to debate his detractors, even in the limited space allowed by television interviews on clearly slanted television news programs, or as James Cone (1995, p. 84) notes, he "eagerly awaited a time when he could look directly at Whites—with fire in his eyes and accusatory right index finger pointing—and tell them what he thought about their Western civilization." Although footage of Malcolm shown in the documentary show all the righteous indignation Cone describes, and while his eyes do hold that fire in the interview he is granted on the program, Malcolm also displays the unflappable personality in the heat of White racism that has become legendary, and one of the best manifestations of "cool" that anyone will ever see.

Contrasted with Elijah Muhammad, Malcolm X relished his role as a formidable adversary to his opponents, and loved to debate them, whether they were Christian pastors, intellectuals, or journalists. He did so calmly, sometimes playfully, often deadly seriously, all with the intent to not only show the weakness of their arguments and the immorality of their views, but to "flip the script" in the sense that Keith Gilyard intended in his book of the same title—to recast the terms of the debate and challenge those who (unintentionally or intentionally) helped to uphold a racist status quo to defend their hostility to the idea of justice for African Americans and/or their gradualist approaches toward achieving that justice. As the documentary shows, Malcolm X, Elijah Muhammad, and the many other guests on the program were required to defend not only their politics or ideologies, but their right to a place in public dialogue, even their existence at times, before being allowed to make any argument they might pursue. Malcolm X's rhetorical genius often lay in his ability to force his opponents onto the defensive, to justify their ideological assumptions before they could attack his positions. Cone celebrates this ability of Malcolm's by comparing him to the consummate boxer, knowing exactly when and how to feint, bob, weave, jab, counter, and deliver the knockout blow to such an extent that many Black integrationists refused to debate him or engage him publicly.

Malcolm X displayed this ability on television, as well—even when the balance of control was so much in Wallace's and the news network's favor. While Lomax's presence as the interviewer and his handling of sensitive lines of questioning helped to shift this balance, any such shift was completely negated by Mike Wallace's blatant editorializing and his constant control over the series as the final arbiter of all that was said. When Lomax asks Malcolm about the continual references to White people as devils and snakes in the Nation's teaching, Malcolm holds his ground on those references, but also makes connections with biblical teaching, telling Lomax that the commentary on Whites is figurative. Lomax follows up on this question asking if Malcolm means to say that all White people are evil, whether there is a White person who is capable of good. The subtlety of Malcolm's rhetoric might be lost on many people at this moment, as it was certainly lost on journalists like Wallace. In response to the question, Malcolm X continues the charge that White people collectively are "evil," but does not say that there is no White person capable of good or that they are all evil. Rather, he points out that the judgment of a nation, a society, a culture has to be rooted in its history, challenging the history of White supremacy directly: "history is best qualified to reward all research, and we don't have any historic example where we have found that they have, collectively, as a people, done good." One of the segments of the documentary presented with considerable alarm the fact that the Black Muslims had chartered independent parochial schools, schools that ostensibly programmed anti-White racism into children from the age of four onward. When asked about the schools and what they taught, again Malcolm focuses not on Whites, but on healthy self respect for African Americans and undermining the cultural impact of White supremacy: "we teach them the same things they would be taught ordinarily in school, minus the little Black Sambo story and things that were taught to you and me when we were coming up, to breed that inferiority complex in us." The sparring continues when Lomax asks what Black children are taught about White Americans, redefining both the devil and hell. In response the question "do you teach them what you have just said to me, that the White man is the symbol of evil?" "You can go to any Muslim child and ask him where is hell or who is the devil, and he wouldn't tell you that hell is down in the ground or that the devil is something invisible that he can't see. He will tell you that hell is right where he is catching it and he will tell you that the one who is responsible for him having received this hell is the devil." By telling Lomax and the program's viewers that the Black Muslim's language is symbolic and that their focus is on calling Black people's attention to the nature of racism in America rather than on White people themselves, he presents Elijah Muhammad's views unequivocally while at the same time telling those viewers that the invective is the equivalent of selling theological wolf tickets, and demonstrating a complexity in those views rarely credited to Muhammad or Malcolm X because of the alarm those wolf tickets caused. He managed all of this in a context where Lomax's line of questioning (no matter how thoughtfully delivered) and its total focus on the accusation that the Black Muslims were hatemongers allowed little room for such carefully developed argumentation. This rhetorical deftness is even more remarkable when one considers that it was achieved on a program that was thoroughly biased against the organization from the start produced and aired in a television news environment where African Americans had absolutely no access or control. This environment was one in which Lomax was silenced not only by the role of newsman as God that Mike Wallace played as the arbiter of truth in every segment aired, but in which the selection of the most sensationalist footage possible, the all-White camera crews that attended Lomax, and African American leaders and activists were demanded to censure Muhammad, Malcolm, the Nation of Islam, and Black nationalist ideology altogether. Given the acclaim John F. Kennedy is often given as the first American president to truly master television as a communicative medium, one can only wonder what he would have done in a televised policy debate with Malcolm X, either before or after his embarrassment of Richard Nixon and eventual election as president.

I make no pretense toward a definitive reading of either Malcolm X or Martin Luther King's use of television. Instead, I want to suggest how a change in the analytic tools we bring to bear on the study of African American rhetoric can allow us to do what we have always done, but better. Such a change can take us to texts and performances that have not received the same attention as Malcolm and King have. It can also lead to new insights on how activists managed to "jack" access through whatever means possible, insights that can lead to strategies for doing the same with new technologies, helping rhetors can acquire far more control over messages than we have had at other times in our history.

King's and X's technological awareness went well beyond the mastery of individual tools, however. In attempting to imagine what he calls an "emancipatory composition," Bradford Stull offers a reorientation of the discipline's goals based on the African American jeremiad, saying that it

> Is radically theopolitical, on the one hand, because as it seeks to set free the captives and give sight to the blind, it roots itself in the foundational theological and political language of the American experience. It is radically theopolitical, on the other hand, because as it does so it calls into question this language, and thus the American experience itself. Hence it is at once conservative and extreme. Emancipatory composition [like the Black jeremiad] would simultaneously dedicate itself to the study of the foundational vocabulary of the nation while embracing what Kenneth Burke calls "the comic attitude," an attitude that would allow it to subvert this vocabulary in order that its telos might be reached: the emancipation of the oppressed, and thereby, as Martin Luther King would say, the oppressors. (p. 3)

Stull offers broad interpretations of ways W.E.B. DuBois, Malcolm X, and Martin Luther King, Jr. can contribute to emancipatory composition arguing, in essence, that these three jeremiahs teach us that the choices between skill drill and kill or critical awareness, between Booker T. Washington and DuBois, between basic access and transformation is a false choice. The Black jeremiad, whose formal properties and use I treat more fully in chapter 5, is about achieving both access and transformation, although they are viewed as representing widely varying, even opposed ideological views. As I argued in my look at the "Falling Through the Net" reports and our own discipline's literature about technology access in the previous chapter, this pursuit of a transformative access stands at great odds with the underlying rhetoric of the Digital Divide in its goal of mere physical, or material access to technologies. Malcolm X and Martin Luther King, Jr. both had far more expansive notions of access to individual technologies like the franchise and the broader role American technological advancement plays in upholding racism. The next two sections of this chapter take up these complex understandings of technologies and two elements of the access puzzle. I use Malcolm X's classic speech "The Ballot or the Bullet" to show the wide range of literacies that make up what I identified as functional access, and King's speech "Remaining Awake Through A Great Revolution" to examine a broader critical access that we must develop if we hope to make meaningful interventions into the ways they operate in our society.

MALCOLM'S BALLOT AND THE LITERACIES OF FUNCTIONAL ACCESS

"Well, I am one who doesn't believe in deluding myself. I'm not going to sit at your table and watch you eat, with nothing on my plate, and call myself a diner. Sitting at the table doesn't make you a diner, unless you eat some of what's on that plate." Very early in his appearance at Cleveland's Cory United Methodist Church in April 1964, Malcolm X described exactly what was at stake in stalled congressional talks on a civil rights bill, the major conflict in which was legal protection of African Americans' access to the franchise. What was at stake in that bill was not just the right to vote, of course, but the social, political, and economic grounds on which African Americans would participate in American society. Those stakes would depend, as Malcolm well knew, on the definitions of access, of the right to vote, operative in any legislation that would be enacted. In other words, this access could not merely be a government pronouncement that the polls would now be open to Black people and the establishment of polling places in Black neighborhoods. We know now how true this is given the events of the 2000 presidential election, and all of the work organizations like the ACLU have done in order to ensure more protections in preparation for the 2004 elections. In this context, Malcolm's speech still holds many lessons for us—not only for our continued battles for meaningful access to the franchise, but for the pursuit of access to digital technologies as well. In this speech, Malcolm's signifying move on the old saying "a place at the table" departs in important ways from his view of the Civil Rights movement at the time, as well as from the Civil Rights movement's arguments for access to the franchise.

Before taking up the analysis of "The Ballot or the Bullet" that follows, some context is in order. For all of the arguments that take place pitting Martin versus Malcolm, the Civil Rights movement versus the Black Power movement, access to American society versus the desire to fundamentally change it, these binaries simply do not hold, no matter how real the differences in the approaches of these activists were. Manning Marable and Leith Mullings (2000) note the overlapping of these different traditions in the introduction to their anthology *Let Nobody Turn Us Around: Voices of Resistance, Reform, and Renewal*:

> integrationism, nationalism, and transformation are not mutually exclusive, but are in fact, broad, overlapping traditions. Throughout the twentieth century, theses tendencies have been present, to varying degrees, in virtually every major, mass movement in which Black people have been engaged, from the desegregationist campaigns of the 1950s to the anti-apartheid mobilization of the 1980s. Though some organizations and individuals may have exemplified one tendency or the other, organizations and movements usually displayed a spectrum of views. (p. xxi)

An example with which to contrast Malcolm X's thorough understanding of the workings of the technology of the franchise will highlight Marable and Mullings' point as well as my argument for a transformative access as the ONE, or a unifying ideal in a Black digital rhetoric. As mild, tempered, or even conservative as Martin King often seemed when contrasted with Malcolm X, his vision was nearly always one of transformation. As early as 1957 the localized goals of the Montgomery Bus Boycotts two years earlier and their focus on integrating public transit became broader. The title of King's keynote address for the Prayer Pilgrimage for Freedom announces this shift: "Give Us the Ballot—We Will Transform the South." The speech is at once a lament that more progress has not been made in ending segregation since the Brown decision, and a declaration that only full enfranchisement of African Americans will make the Brown decision meaningful.

After noting the political and criminal opposition to Brown that sparked everything from "crippling economic reprisals to the tragic reign of violence and terror," (1991, p. 197) King argues that the right to vote for Black people will necessarily force dramatic change:

> Give us the ballot and we will not longer have to worry the federal government about our basic rights.
>
> Give us the ballot and we will no longer plead to the federal government for passage of an anti-lynching law; we will by the power of our vote write

> the law on the statute books of the southern states and bring an end to the dastardly acts of hooded perpetrators of violence.
>
> Give us the ballot and we will transform the salient misdeeds of bloodthirsty mobs into the calculated good deeds of orderly citizens.
>
> Give us the ballot and we will fill our legislative halls with men of good will, and send to the sacred halls of Congress men who will not sign a Southern Manifesto, because of their devotion to the manifesto of justice.
>
> Give us the ballot and we will place judges on the benches of the South who will "do justly and love mercy," and we will place at the head of the southern states governors who have felt not only the tang of the human, but the glow of the divine.
>
> Give us the ballot, and we will quietly and nonviolently, without rancor or bitterness, implement the Supreme Court's decision of May 17, 1954.

The intensity rising throughout the list of King's repetitions and his announcements of all African Americans will do if granted the franchise reveals a paradox. King's vision for the nation is clearly a transformed one, but his faith in the ability of the ballot to bring all of this about betrays a naivete reminiscent of the faith educators, politicians, and computer marketing campaigns placed in computers and the Internet to solve all our society's problems.

Malcolm suffered from no such delusions. He announces from the beginning that African Americans have been denied even the most basic access to the political machinery that the vote sets in motion "Sitting at the table doesn't make you a diner, unless you eat some of what's on that plate." Malcolm X continues his critique of segregation, confirming for many the labels of separatist with which he was tagged: "I'm not a politician, not even a student of politics; in fact, I'm not a student of much of anything. I'm not a Democrat, I'm not a Republican, and I don't even consider myself an American. If you and I were Americans, there'd be no problem." He continues reflecting the irony of European immigrants' relatively easy absorption into the mainstream of White privilege: "Polacks are already Americans; the Italian refugees are already Americans. Everything that came out of Europe, every blue-eyed thing, is already an American. And as long as you and I have been over here, we aren't Americans yet" (1989, p. 427).

After the first sentences of the speech, it is apparent that Malcolm X's comment is not one reflecting voluntary withdrawal from American systems and society but the longstanding denial of full citizenship to Black people. He then proceeds to argue that African American experience in the United States shows the hypocrisy of the notion of citizenship by birth. Citizenship and access to all it entails is granted and withheld by processes that begin with the vote. As Malcolm continues, "being born here in America doesn't make you an American. Why, if birth made you American, you wouldn't need any legislation, you wouldn't need any amendments to the Constitution, you wouldn't be faced with civil rights filibustering in Washington, D.C. right now. They don't have to pass any civil rights legislation to make a Polack an American" (p. 427).

Malcolm X's speech is a rhetorical tour de force for many reasons; one of the often overlooked reasons is the detailed understanding it shows—and provides—its listeners of how the vote really works in the American political system. The speech shows both a thorough grasp of the literacies one needs to work the electoral process and a critical perspective of when it should be used and when it should not.

The threat promised by the speech's title and delivered by its text is that a new day has emerged: whether through the votes of a fully enfranchised African American population, or through the bloodshed of violent revolution, America will be changed. While many people were alarmed by the rhetorical threat of violence used by Malcolm X and many 1960s militants, the real threat to the nation's racist status quo lied in a Black electorate who not only had the ballot, but knew what to do with it, and went out and did it:

> These 22 million people are waking up. Their eyes are coming open. They're beginning to see what they used to only look at. They're becoming politically mature. They are realizing that there are new political trends from coast to coast. As they see these new political trends, it's possible for them to see that every time there's an election the races are so close they have to recount. They had to recount in Massachusetts to see who was going to be governor, it was so close. It was the same way in Rhode Island, in Minnesota, and in many other parts of the country. And the same with Kennedy and Nixon when they ran for president. It was so close they had to count all over again. Well, what does that mean? It means that when White people are evenly divided, and Black people have a bloc of votes of their own, it is left up to them to determine who's going to sit in the White House and who's going to be in the dog house. (Ballot or the Bullet, 1965)

The awareness Malcolm displays of voting trends in state and national elections is valuable, but not in and of itself. Its importance lies in the ability to move from that general sense of the political situation in 1964 to an assessment of exactly where the possibilities for the power of Black access to the ballot could lie. African Americans have learned this lesson well, but have had those lessons countered by White politicians, Democrat, and Republican, by their efforts to continually dilute Black voting strength through the gerrymandering of political districts along racial lines.

Malcolm X notes these practices himself, with an eye toward changing African American access to the ballot:

> I was in Washington, D.C. a week ago Thursday, when they were debating whether or not they should let the bill come onto the floor. And in the back

> room where the Senate meets, there's a huge map of the United States, and on that map it shows the location of Negroes throughout the country. And it shows that the Southern section of the country, the states that are most heavily concentrated with Negroes, are the ones that have the senators and congressmen standing up filibustering and doing all other kinds of trickery to keep the Negro from being able to vote. This is pitiful. But it's not pitiful for us any longer; it's actually pitiful for the White man, because soon now, as the Negro awakens a little more and sees the vise that he's in, sees the bag that he's in, sees the real game that he's in, then the Negro's going to develop a new tactic. (p. 29)

That new tactic is knowledge of when and how to use the ballot toward African American interests, to realize "what the ballot is for; what we're supposed to get when we cast a ballot; and that if we don't cast a ballot, it's going to end up a situation where we're going to have to cast a bullet." In the speech, Malcolm X also challenges his listeners to take up a broader understanding of the technology of the franchise. He makes clear that it is not only the artifact (the physical ballots and the machines and systems used for casting them). It is also the system of people and processes in which it operates—the legislators, executives, and judges elected by those ballots, the laws they pass, enforce and interpret, the budgets they have discretionary power over, the individuals they are able to hire and fire, the police (along with their guns, dogs, hoses, and clubs) used to uphold them. Knowing how to use the vote requires an understanding of all these interrelated agents and processes, and the specific means voters have for using or changing them, including appeals to the United Nations, revolutionary violence, and ultimate, final separation from the nation and its voting technologies.

More than all of this, however, Malcolm X's speech launches heated critique of the ways voting systems have been denied Black people and used to reinforce their reduction to second-class citizens. He also critiques the ways African Americans have failed to understand or use the power available to them. The problem of racism is not merely the fault of violent segregationists or those who hold prejudiced views of Black people, Malcolm argues, but of those who would scorn the actions of the Ku Klux Klan yet continue to use governmental systems to uphold unequal power relations: "you and I in America are not faced with a segregationist conspiracy, we're faced with a government conspiracy. Everyone who's filibustering in Washington, D.C., is a congressman—that's the government. You don't have anybody putting blocks in your path but people who are a part of the government." The point here is that Malcolm X is critical of the system and its agents who would argue that they are not directly responsible: "it is the government itself, the government of America, that is responsible for the oppression and exploitation and degradation of Black people in this country. And you should drop it in their lap." To reinforce how thoroughly he believed that the government's failure was a failure of the entire system, Malcolm adds "this so-called democracy has failed the Negro. And all these White liberals have definitely failed the Negro."

The systematic nature of this critique stands in stark contrast to King's utter faith in the change that Black voting rights can initiate, even as Malcolm X agrees with King that informed access to the vote can and must change the political process. It also stands in stark contrast to analyses of racial exclusions that seek to blame African Americans for the intransigence of American racism—although Malcolm refuses to only critique the system. He calls out Black voters on how they use the vote and their orientation toward the larger system as well: "your vote, your dumb vote, your ignorant vote, your wasted vote put in an administration in Washington, D.C. that has seen fit to pass every kind of legislation imaginable, saving you until last, and then filibustering on top of that." He then shows even scorn for the patient approach of civil rights leaders and activists, saying

> and your and my leaders have the audacity to run around clapping their hands and talk about how much progress we're making. And what a good president we have. If he wasn't good in Texas, he can't be good in Washington, D.C. Because Texas is a lynch state. If is in the same breath as Mississippi, no different; only they lynch you in Texas with a Texas accent ... And these Negro leaders have the audacity to go and have some coffee in the White House with a Texan. (p. 27)

The overall lesson that Malcolm attempts to teach in this speech is that users of any technology have to know both how to use and when to refuse any and all elements in that system. The informed refusal that a critical access provides is just as important as the informed use that the literacies of functional access allow. Like King, Malcolm envisions a nation transformed by African American access to the ballot, a ballot fully protected by all the means at federal and state governments' disposal, and in the specific uses to which newly enfranchised voters would put their votes.

Where Malcolm X's speech shows a thorough functional knowledge and critical awareness of the particular technology of the ballot and the larger political system of which that ballot is a part, Martin Luther King, Jr.'s "Remaining Awake Through a Great Revolution" addresses larger questions concerning the roles of technologies in human life. In the speech, King insists that the test of technological advancement is not increased efficiency or profit, but the degree to which those technologies help to end human suffering and injustice. Religion scholar James Cone documents the ways African American preachers transformed interpretations of the Bible and Christianity more broadly with a "liberation theology" that placed Christ squarely on the side of the oppressed. The purpose of the Black church, then, became about actively seeking the liberation of African Americans, from the individual spirit to the congregation to the nation to Black people worldwide.

The radical achievement of liberation theology is in its ability to take what was a tool of enslavement and racism and use it aggressively to disrupt the status quo. This is the kind of reinterpretation King offers of the role of technologies. In it, he establishes his own activism and African American struggle as examples of technology theorist Andrew Feenberg's "third way," arguing that neither time nor technology are positive or negative forces in and of themselves. Instead, they reflect the ideological commitments societies and individuals impose on them. What this means is that one must call attention to the systemic problems in the ways technologies have been used to further oppression—but also that there are always ways to resist, that there is always agency in the individual and in the society. This argument allows King to maintain hope in the world even as he confronts the magnitude of human suffering on a global scale and constant reminders that much of the suffering he sees is connected to the intransigence of racism and racialized patterns of exclusion.

In the text of this speech, delivered at the National Cathedral on March 31, 1968, just days before his death, King uses the story of Rip Van Winkle to warn his listeners of what is at stake in failing to attend to the problems involved in technological advancement. The biblical basis for King's parallel hope in the possibility of transformation is found in the 16th chapter of the book of Revelation: "Behold, I make all things new; former things are passed away." The lesson King draws from Washington Irving's story is not merely that Rip slept a long time, but that he awoke to an entirely different world:

> there is another point in that little story that is entirely overlooked. It was the sign in the end, from which Rip went up in the mountain for his long sleep .... When Rip Van Winkle looked up at the picture of George Washington—and looking at the picture he was amazed—he was completely lost. He knew not who he was. (p. 268)

King's lesson about these dangers is simultaneously addressed to African Americans and their allies engaged in civil rights struggle, and the larger society. The all risk sleeping through what King identifies as a "triple revolution." The elements of this revolution seem to be in conflict: "a technological revolution, with the impact of automation and cybernation; then there is a revolution in weaponry, with the emergence of atomic and nuclear weapons of warfare; then there is a human rights revolution, with the freedom explosion that is taking place all over the world. And there is *still* the voice crying through the vista of time saying, "behold, I make all things new" (p. 269). The conflict inherent in this triple revolution is the overlapping of the first two elements in their ability to stem the last—that all of the machines and weapons held by those in power can be used to neutralize the yearnings of those colonized nations and oppressed groups for powers of their own. Those fighting on behalf of oppressed peoples in the world face great danger if they fail to consider the ramifications of those technologies; those in power in the United States and other colonial powers face damnation just as sure if they sleep through the revolution of oppressed people refusing to live with that oppression. So the challenge of transformation is directed at both ends: systems of power must change and those who want power must change.

The answer King offers to these challenges is one that many technology proponents have offered—the idea of the world as a global community or neighborhood. He tells his listeners: "first we are challenged to develop a world perspective. No individual can live alone, no nation can live alone, and anyone who feels he can live alone is sleeping through a great revolution. The world we live in is geographically one" (p. 269). King's globalism is far different, however, than that of the governments and companies that have appropriated his voice in their service. "The challenge we face today is to make it one in terms of brotherhood. Now it is true that the geographical oneness of this age has come into being to a large extent through modern man's scientific ingenuity. Modern man has through his scientific genius been able to dwarf distance and place time in chains" (p. 269).

The brotherhood King posits as necessary to our survival is different because it depends on a mutual responsibility that demands the correction of past injustices, with American racism chief among them. This is what cable news channels fail to grasp when they play their favorite clips from the famous "I Have a Dream" speech, and what neoconservatives either fail to understand or ignore when they sample his words for a song he wouldn't sing. The difference between King's global vision and corporate or governmental globalisms, between a global neighborhood and a global brotherhood, is the same as the difference between actively including those who have been denied technological access and casually placing a few computers in a few libraries and arguing that it is the responsibility of those who have been denied to find their own way.

One cannot escape the responsibility King places on those in power. He embraces technologies and those who have created them, but demands change:

> through our scientific and technological genius, we have made of this world a neighborhood and yet we have not had the ethical commitment to make of it a brotherhood. But somehow, and in some way, we have got to do this. We must learn to live together as brothers or we will all perish together as fools. We are tied together in the single garment of destiny, caught in an inescapable network of mutuality. And whatever affects one directly affects all indirectly. For some strange reason I can never be what I ought to be until you are what you ought to be. (p. 269)

The clarity of King's challenge is unmistakable. It stands in direct opposition to histories of White (and later Black) flight, dependence on a prison industrial complex to remove Black men and women from the labor pool, and "tolerance" and "diversity" as ways of responding to the history of racism.

After moving from his general argument for mutual responsibility at the heart of a global brotherhood, King directly attacks American racism as "a way of life for the vast majority of White Americans, spoken and unspoken, acknowledged and denied, subtle and sometimes not so subtle," a disease which "permeates and poisons a whole body politic" (p. 271). King then moves to debunk the twin myths that prevent much of the nation from working to heal this disease—that time will heal all wounds, and that it is the sole responsibility of the victims of racism to find their own place in the society. The arguments he makes in both places, while not directly invoking technology access, still speak to the debates about the Digital Divide I detailed in chapter two.

The first myth King identifies, "the myth of time ... the notion that only time can solve the problem of racial injustice" can be read as a direct response to what is known in technology circles as diffusion theory. The "diffusion of innovations" as a theory states that technologies will work their way through a society in relatively predictable terms, unrelated to patterns of injustice or exclusion, but connected instead to people's acceptance of a given technology and their view of it as relevant to their lives. Thus time is the important variable, and not exclusion—any problems in access that do occur will eventually work themselves out. In essence, it is the same myth, the myth that "time can solve the problem of racial injustice," that King addressed repeatedly throughout his career, especially when addressing liberal and moderate Whites. King's answer in this speech is very similar to what Andrew Feenberg's (1995b) understanding of technology in his essay "Subversive Rationalization: Technology, Power, and Democracy." Table 3.1 places quotations from each side by side in order to illustrate the similarities.

Although King asserts that time is "neutral," he means very nearly what Feenberg means—that belief in time or nature or technology as forces that will somehow magically make the world right is mistaken. The technological transformation that King argues for, that we find ways to use our technologies to make the world a brotherhood with an ethic of mutual responsibility, is "intrinsically social," just as Feenberg argues. King's speech then, is an argument for the kind of social outlook that can transform the development and impact of the particular technologies that make up our society.

King's attack on the ideas that are an "ally of the primitive forces of social stagnation" comes by way of an extended reflection on the "bootstrap philosophy" that is the legacy of Booker T. Washington and the many people of all races who agreed with him through the years. King's engagement and dismantling of this idea are especially important, because Washington's call to Black people to "cast down your buckets where you are" was part of a complex response to the industrialization, through machining technologies, of 19th century America. Washington's ideology was a very persuasive set of arguments that attempted to help Black people acquire the tools and capital that would allow them a place in American society after the failed American Reconstruction.

The second myth that King addresses is "a kind of overreliance on the bootstrap philosophy. There are those who still feel that if the Negro is to rise out of poverty, if the Negro is to rise out of the slum conditions, if he is to rise out of discrimination and segregation, he must do it all by himself." King's response to this myth rests not only on the particularity of the African American experience of slavery, but on the many ways American society sacrificed to assist other groups of people in their pursuit of access to credit, capital, knowledge, and technology. Those who insist

TABLE 3.1

King in "Remaining Awake Through a Great Revolution"	*Feenberg in "Subversive Rationalization: Technology, Power, and Democracy"*
One is the myth of time. It is the notion that only time can solve the problem of racial injustice. And there are those who often sincerely say to the Negro and his allies in the White community, "Why don't you slow up? Stop pushing things so fast. Only time can solve the problem. And if you will just be nice and patient and continue to pray, in a hundred or two hundred years the problem will work itself out. There is an answer to that myth. It is that time is neutral. It can be used either constructively or destructively. And I am sorry to say this morning that I am absolutely convinced that the forces of ill will in our nation, the extreme rightists of our nation—the people on the wrong side—have used time much more effectively than the forces of goodwill. Not merely for the vitriolic words and the violent actions of the bad people, but for the appalling silence and indifference of the good people who sit around and say "wait on time." Somewhere we must come to see that human progress never rolls in on the wheels of inevitability. It comes through the tireless efforts and persistent work of dedicated individuals who are willing to be co-workers with God. And without this hard work, time itself becomes the ally of the primitive forces of social stagnation. So we must help time and realize that the time is always ripe to do right. (pp. 270–271)	So far as decisions affecting our daily lives are concerned, political democracy is largely overshadowed by the enormous power wielded by the masters of technological systems: corporate and military leaders, and professional associations of physicians and engineers. They have far more to do with control over patterns of urban growth, the design of dwellings and transportation systems, the selection of innovations, our experience as employees, patients and consumers, than all of the governmental institutions of our society put together … No doubt modern technology lends itself to authoritarian administration, but in a different social context it could just as well be operated democratically. In what follows I will argue for a qualified version of this … position, somewhat different from the usual Marxist and social-democratic formulations. The qualification concerns the role of technology, which I see as neither determining, nor neutral. I will argue that modern forms of hegemony are based on the technical mediation of a variety of social activities, whether it be production or medicine, education or the military, and that consistently, the democratization of our society requires radical technical as well as political change. [Technology] is not just the rational control of nature; both its development and impact are intrinsically social. (pp. 3–4)

that Black people alone are responsible for their inclusion or exclusion from American society, in King's view

> never stop to realize that no other ethnic group has been a slave on American soil. The people who say this never stop to realize that the nation made the Black man's color a stigma. But beyond this they never stop to realize the debt they owe a people who were kept in slavery two hundred and forty years. In 1863 the Negro was told he was free as a re-

> sult of the Emancipation Proclamation being signed by Abraham Lincoln. But he was not given any land to make that freedom meaningful. It was something like keeping a person in prison for a number of years and suddenly discovering that the person is not guilty of the crime for which he was convicted. And you just go up to him and say, "now you are free," but you don't give him any bus fare to get to town. You don't give him any money to get some clothes to put on his back or to get on his feet again in life. Every court of jurisprudence would rise up against this, and yet this is the very thing our nation did to the Black man. It simply said, "you're free," and left him there penniless, illiterate, not knowing what to do. And the irony of it all is that at the same time the nation failed to do anything for the Black man, through an act of Congress, it was giving away millions of acres of land in the West and Midwest. Which meant it was willing to undergird its white peasants from Europe with an economic floor. But not only did it give the land, it built land-grant colleges to teach them how to farm. Not only that, it provided country agents to further their expertise in farming. Not only that, as the years unfolded it provided low interest rate loans so they could mechanize their farms. And to this day thousands of persons are receiving millions of dollars in federal subsidies every year not to farm. And these are often the very people who will tell Negroes that they must lift themselves up by their own bootstraps. It's all right to tell a man to lift himself up by his own bootstraps, but it is a cruel jest to say to a bootless man that he ought to lift himself up by his own bootstraps. (1991, p. 271)

This is a lengthy quote from King, to be sure, but it is justified when considered in light of the Dr. King we imagine now, sanitized from a safe distance of almost 40 years after his death. We find ourselves comfortable with, even taking solace in his vision of an America that recognizes people's gifts and holds hands and sings "KumBaYah" together, but we conveniently ignore his critique of the structural basis of American racism and his insistence that we are responsible for each other. It is also justified when considered in light of the current moment of racial recalcitrance in which we live, when even the very minimal protections provided by civil rights era legislation and government policy have been systematically eroded and attacked as "reverse racism."

There are several reasons why I chose to analyze Martin and Malcolm as collectively embodying a model for a Black digital ethos. The first is that the Digital Divide is a rhetorical problem. The persistence of the Divide and all of its attending ironies is the failure of current social and technology activists to develop the arguments and the language to promote meaningful material, functional, experiential, and critical access to technologies in a larger project of the transformation of those technologies and the larger society. It is also the failure of those who are reaching for that language, for those arguments to find effective means for presenting them. It is also our failure to use the means we have—print, visual, and electronic—effectively. A Black digital ethos is the combination of all these abilities Martin Luther King and Malcolm X demonstrated, rooted in African American traditions, committed a larger project of transformation to make some real difference in the lives of Black people, however they define that difference. It is rhetorical excellence based on both a mastery of skills and broader critical awareness. It is an understanding of the workings of individual technologies and the larger networks of power in which they operate. It is a vision for the role of technology in its most general sense in the lives of Black people and in a common humanity. It is an ability to stage the event, manage the interview, give the speech, create the weblog, bring the television cameras, get on the radio, preach at the church, or whatever else is necessary, in whatever forum, with whatever tools, to carry one's message to the people regardless of the differences of power that might exist, regardless of the knowledges, needs and assumptions audiences bring with them. A Black digital ethos is the set of attitudes, knowledges, expectations, and commitments we need to bring to our dealings with individual technological tools and to that larger, macrolevel awareness we need. Malcolm and Martin and all the networks of their collaborators, parishioners, opponents begin to show us the way.

4

Taking Black Technology Use Seriously: African American Discursive Traditions in the Digital Underground

The first, and perhaps most important element of a meaningful access is use—more than merely owning or being close to some particular technology, people must actually use it, and develop the skills and approaches to using it that are relevant to their lives. Unfortunately, we know almost nothing about the uses to which African Americans put digital technologies or the processes by which they develop the skills, abilities, and approaches that will enable them to use computers, the Internet, or any other related tool or process in culturally relevant, individually meaningful ways.

This chapter raises the issue of how African Americans use computer-related writing technologies, and tries to answer the specific question of how Black computer users engage African American language and discourse patterns in online spaces. I argue here that despite the prevalence of work on the oral elements of African American language and discourse, and the dominance of early cyberspace theory that dismissed race and culture as irrelevant online, at least in one space, African American community Web site BlackPlanet, African American language and discourse live, and even thrive in online spaces. The strength of the presence of Black language and discourse online speaks not only to the richness of Black linguistic and discursive traditions, but also to the ways African American technology users can change technologies and make them relevant through the uses to which they put them—even when those technologies are not Black owned or controlled. Black participation on the Web site also begins to show the ways cyberspace can serve as a cultural underground that counters the surveillance and censorship that always seem to accompany the presence of African Americans speaking, writing, and designing in more public spaces—spaces that seem to consistently say to them that no matter what traditions they might bring to the classroom, the workplace, or to technologies—these spaces (and the written English that accompanies them) are, and will continue to be White by definition.

As much attention as people have paid to the structures, features, and functions of African American varieties of English—whether those varieties are referred to as African American Vernacular English, Ebonics, a creole, or a complete language that is a member of the Niger-Congo language family, the great preponderance of that attention has focused on the "oral" elements of those language traditions. In fact, one might say that just as African American rhetorical scholarship has tended to focus on oratory or other orally delivered texts like song lyrics, African American language study has taken African and African-derived oral traditions as givens, at times seeming to concede written language to the domain of White culture. Of course, there are many African American language scholars who do not make such concessions, but the broader body of research would seem to make it for them.

Although there has been much work done examining the ways African American language patterns are manifest in written texts, the fact that the focus of much of that study and most of the theory of those language patterns has been on oral language production is clear. Titles from two of the more prominent books on Black English tell the story: John and Russell Rickford's 2000 book, *Spoken Soul: The Story of Black English*, and Geneva Smitherman's compilation, also from 2000, *Talkin that Talk: Language, Culture, and Education in African America* begin to show that focus. The Rickfords take their title from the name Claude Brown gives Black English in an interview on language, and they point to several reasons for borrowing Brown's phrase: a desire to stay rooted in the vernacular tradition of African American English, its links to other forms of cultural production like music, and "the fact ... that most African Americans *do* talk differently from Whites and Americans of other ethnic groups, or at least most of us can when we want to. And the fact is that most Americans, Black and White, know this to be true" (p. 4).

Smitherman's focus on orality in *Talkin that Talk* and her classic first book *Talkin and Testifyin: The Language of Black America* (1986) exists for similar reasons, though she also pushes the discussion further to account for distinctively Black rhetorical strategies and discourse features that, although rooted in Black oral traditions, are manifest in written texts as well. Her work also connects these specific discursive patterns and rhetorical features to African American epistemologies, or world views. As Smitherman shows in the essay "How I Got Ovuh: African World View and African American Oral Tradition," the point is not simply a linguistic one. In other words, simply arguing that African American English is, obviously (thought not yet for many teachers), rule governed and systematic, and just as valid a organ for expressing the thoughts, ideas, passions, joys, pains, aspirations, and struggles of a people as any other is no longer news for those who study the language variety. A whole legion of linguists have been on that case since Lorenzo Dow Turner more than 50 years ago.

Smitherman goes beyond this point to show how Black English, as expressed through its oral traditions, represents distinctively African American worldviews:

> the oral tradition has served as a fundamental vehicle for gittin ovuh. That tradition preserves the Afro-American heritage and reflects the collective spirit of the race. Through song, story, folk sayings, and rich verbal interplay among everyday people, lessons and precepts about life and survival are handed down from generation to generation. (p. 199)

The continued focus of many on the oral in Black English, then, is not a resignation that written English is somehow the exclusive domain of Whites—Smitherman herself offers masterful analysis of a wide array of written texts—but a matter of remaining true to the roots of the language, no matter what forms it might take now. Maintaining that focus is also an act of self-determination, of resistance, of keeping oppositional identities and worldviews alive, refusing to allow melting pot ideologies to continue to demand that Black people assimilate to White notions of language and identity as the cost for access to economic goods or a public voice in American society.

THE "UNDERGROUND" IN AFRICAN AMERICAN CULTURE: THRIVING AND EXTINCT

The underground in Black life that the oral tradition still dominates has become so popular as to almost no longer be metaphorically underground. By "underground," I refer to those spaces and cultural practices that exist away from the policing gaze of mainstream culture. I use it to call attention to those spaces that, whether initially created as ways of resisting the segregation and racism that prevented Black people from participating in mainstream American society (and usually punished them for trying), or as refuges from the constant struggle in trying to swim that stream, are dominated by Black cultural norms, Black worldviews, and Black language practices. The most well-known of these sites, and the source of the concept of a Black underground is the Underground Railroad, a network of people, organizations, communication technologies, and language practices that Black people used to resist and escape the horrors of slavery. The railroad and other spaces and practices are termed *underground* because it operates when and where others fail to notice, although often in plain view.

This tension, between maintaining distinct cultural practices, often in plain view of others, is important to bear in mind because many recent popular culture texts seem to suggest that there is no underground anymore. The traditional sites of the underground have all been exposed—the barbershop and beauty salon, the nightclub, the church, standup comedy, the radio show, the studio, the basement.

Hardly a Black sitcom or movie appears on television or onscreen without pointing to these sites: Nipsey's Bar, and the radio show that employs the lead character on "Martin," BET and HBO comedy shows "Comic View," "Def Comedy Jam," and even music video shows like "Rap City: Tha Bassment." No matter how much exposure spaces like these receive, although, the underground still exists because the exclusions that foster such spaces still exist. It also still thrives because the underground is much more than the physical spaces of the church or barbershop or studio. The underground is also the specific discursive practices that determine who gets in and who does not, from the posters in Black communities advertising concerts to the "grapevine" that Patricia Turner examines in her book on rumor in African American communities. The underground is the particular technologies, tools, processes that make those discursive practices possible. The chitlin circuit of locations comics perform on while attempting to "make it," and mix CDs are two examples of how networks and technologies are sites of underground cultural practices in addition to the physical spaces that receive so much exposure. Technologies, discursive practices, and networks of people all come together with physical spaces to create any particular manifestation of the "underground."

Common to almost all of these sites is the African American oral tradition—not just in the sense of Ebonics (but that too, often), or linguistic notions of communication, but all of the rhetorical and discursive features Smitherman (2000) describes so fully in *Talkin that Talk*. Because these spaces exist outside of the official gaze of schools, workplaces, and governments, those who become part of them truly do have the right to their own language. The presence of such spaces online would mean three things: first, it would be a repudiation of much early cyberspace theory that insisted race is and should be irrelevant online, that it would be made irrelevant by the fluidity inherent in online subjectivities. Second, it would confirm the importance of discursive and rhetorical features that Smitherman links to African oral traditions for the written discourse of African Americans. The implication of this point is that the wide body of research in composition that fails to take into account the power of these traditions and continues to view Black students writers as less prepared than others, merely "writing like they talk" needs to be questioned and ultimately repudiated as antithetical to both the field's stated goals of fostering inclusion in writing instruction and the actual practice of writers. Third, it would show Black people taking ownership of digital spaces and technologies and point to the importance of taking Black users into account in technology user studies.

An examination of African American discourse online, then faces two distinct problems from the outset. The first is the assumption that is incorrectly taken from the work of scholars like the Rickfords, Geneva Smitherman, Keith Gilyard, Elaine Richardson, Clinton Crawford, and many others, that African American English is merely or mostly about spoken language. This assumption often leads teachers and scholars to believe a corollary assumption, namely while Black oral traditions are rich and varied, those traditions are irrelevant to the study of written

English, and therefore writing instruction should be strictly and only about the mastery of standardized English.

Just as many compositionists hold these assumptions that Black traditions are irrelevant to the study of written English, much cyberspace theory also holds that race is, and should be, irrelevant online. Fortunately, some have confronted such theorizing, like Beth Kolko, who argues in her essay "Erasing @race: going White in the interface" that it's not just cyberspace theory or discourse that assumes a White default user, but that this assumption is often programmed right into the interface of online environments like MOOs and MUDs, the older sisters and brothers of current chat spaces. Kolko contends that the elision of race that occurs in these design choices negatively affects communicative possibilities, but also that "the history of online communities demonstrates a dropping out of marked race within cyberspace" (p. 214). This problem is such a significant one, Kolko argues, because "interfaces carry the power to prescribe representative norms and patterns, constructing a self-replicating and exclusionary category of 'ideal' user," a user that is almost always a "definitively White user" (p. 218). The construction of Black people and other people of color as non-technological and therefore irrelevant in the design and construction of technological tools continues even into the era of the Internet, even as those selling new technologies are quick to market a world of multicultural possibility.

TALKIN B(L)ACK TO THE DIVIDE: BLACK AGENCY IN THE JIM CROW CYBER SOUTH

Even if the persistence of exclusion and racism online make cyberspace seem more like the old Jim Crow south than the age of limitless possibilities it was sold as, Black people also have agency online, and there is a critical need to pay attention to the ways African Americans who have crossed the Digital Divide talk b(l)ack—to postmodern theories of race, to ideas about the role of technology in African American life, and to thoughts about how to address problems of systematically differentiated access. In at least one place on the web, a community Web site called BlackPlanet, African American internet users respond to these issues. Race and voice come together to do what David Holmes suggests in "Fighting Back by Writing Black:" race and voice "can be used to map territory, create community, and ensure an ongoing sense of self—and group—affirmation" (p. 65). In the sections that follow, I examine how BlackPlanet's users use both language and technology to these ends, how—in this space, at least—"race as a way of writing affords the best historical, ideological, and pedagogical response" (p. 66) to the Digital Divide. BlackPlanet's users also make important arguments for the roles of play, experimentation, and community building in learning and technology use, offering important cautions against the narratives of crisis that often accompany efforts toward literacy, writing instruction, and technology access.

"Sometimes I wish they didn't even have these things in here. Kids just have no idea how powerful a tool they have in front of them. If we had something like this when we were in school, hell, I would have graduated at 15 or 16." This comment, from a Cleveland Public Library employee, says a great deal about how people who are charged with providing technology access understand technologies, understand access. I worked part-time for the library at this time, filling in at neighborhood branches assisting patrons when the library system was just beginning to make computers and Internet connections available to its patrons. Their approach to this task was to outfit each branch with four to six computers in the hope that patrons would become more technologically literate as a result.

And the patrons responded—especially children and young adults. Librarians and other staff members at many branches became disillusioned, however, because these children and young adults weren't making what they saw as "productive" uses of computers or the Internet. The most popular activities were chatting on sites like BlackPlanet, Yahoo, BlackVoices and others, downloading music lyrics (often the uncensored versions, which really frustrated the staff), and viewing music-related Web sites. Some branches specifically prohibited chatting and downloading music lyrics, and some branches have policies banning these activities at certain hours.

The Cleveland Public Library employee's comment points to several issues, from cultural nostalgia to the frustration librarians and library staff feel about not being able to do more to help young people, to the potential uses students and others might put computers. I include it here to show the ways recreational uses of technologies are often seen as problematic rather than valuable. There are many people for whom a study of African Americans' recreational uses of the Internet would seem odd, or even counterproductive. There would appear to be far more important sites in which to study African American technology use and far more important kinds of online communication to study—e-mail and other workplace documents, college student writing, online student-teacher interaction, political organizations' or grassroots activists' uses of digital technologies. But there is a case to be made that African Americans' recreational uses of the Web are just as important a subject of study as any other, because those uses occur in spaces that are removed from the disciplinary forces of schools, libraries, and other organizations where literacies are taught. Ironically, these recreational uses often occur in libraries and schools.

As genuinely committed as many in these institutions are to providing African Americans and other people of color with meaningful access to digital technologies, that commitment is often spooked by the ghost of Quintilian's "good man speaking well"—the rhetor trained for official or public discourse. Recreational spaces like BlackPlanet allow for a fuller, more organic view of African American rhetorical production: vernacular sites like this provide the opportunity to see what patterns emerge outside the prescriptions used to prepare speakers for public communication. This look for vernacular rhetorical practices on BlackPlanet show users have taken space online to develop the kind of raced voice David Holmes, in connection with Larry Neal, Nellie McKay, Henry Louis Gates, and an entire literary and literate tradition, argue is essential to resistance to, and participation in, unwelcoming or outright racist environments. The heavy use of Black linguistic, discursive, and rhetorical patterns one finds on the site connects those traditions to digital futures, connects access to resistance and transformation, no matter how small the scale.

Before one even examines the site, BlackPlanet's name unmistakably announces the site's intentions. Some recognize it as an appropriation of Public Enemy's 1990 album "Fear of a Black Planet." The name announces that it is not any other hybrid, fluid space online, but rather a separate space; a space where all are welcome to visit or become members, but a distinctively Black space nonetheless. Black people "live" here, and control at least some portion of the virtual space through their uses of the site. This notion of even partial control has to be complicated somewhat: while its expressed intent is to serve African American computer users, and is run by an African American CEO, BlackPlanet is not Black owned. This clearly presents limitations, but these limitations do not entirely negate the importance of the practices users engage in on the site, just as African American musicians, DJs, and club and concertgoers often danced, played, and sang in spaces that were not Black owned. Ownership is clearly important, but the cultural dimensions of language and technology use emerge even through the material relationships that determine ownership of technology companies.

BlackPlanet's potential as an underground site and one that shows the possibility of transformations of individual spaces, then, is a potential that emerges from use rather than ownership, and these uses show the degree to which members have claimed the space as their own. Black discursive conventions are the default, rather than easily dismissed as other. Tricia Rose speaks to the implications of this resetting of the default for our understandings of technologies and rhetorics in a chapter bearing the same name "Fear of a Black Planet: Rap Music and Cultural Politics in the 1990s." For Rose, the resistance inherent in rap music

> involves the contestation over public space, expressive meaning, interpretation, and cultural capital. In short, it is not just what one says, it is where one can say it, how others react to what one says, and whether one has the means with which to command public space. Cultural politics is not simply poetic politics, it is the struggle over context, meaning, and public space. (p. 277)

Just like with early assessments of rap and HipHop culture, much of what we want to assume is just "noise" is connected intricately to ideological, political, and rhetorical struggle. This struggle is exactly what takes place on BlackPlanet.

Individual usernames begin to show this claiming of space: space for an individual identity on the Planet, a connection to African American and Afro-diasporic culture, and public space in broader online and real world discourse. It is sometimes easy to dismiss the importance of online nicknames as merely whimsical, but their importance is clear here on BlackPlanet for several reasons. One of those reasons is that they show the site's members embracing Blackness clearly and openly, and in rich complexity, in clear opposition to theories about computers and cyberspace that assumed fragmentation and "identity tourism" would be the norm, or those theories and technology practices that assumed Whiteness as a default, because after all, "it doesn't matter who you are online." These nicknames make connections to Black musical traditions (2in2Prince, MoreHouseBlues, oldschoolmusicman, kweli, methodwoman); they claim space and authority to speak on political, cultural, and technical subjects (BlackbyDesign, WebDesignTips, liberatedlady); they proclaim Black sexuality as healthy in spite of the power of stereotypes of that sexuality (SensualOne, skillz); they show members' participation in African American organizations (Blackman_06, AKATude, EasternStar); they misbehave and poke fun at assumptions about "good" behavior (legitballin, oldskoolplaya, dpimptress) they resist monolithic beauty standards and identify Black aesthetics as the standards for beauty in this space (bmorelocs, chocolatebeauty, redbonenubian). These names and the names of other members reveal complexity and diversity in notions of exactly what constitutes a Black identity, but all of the users—many overtly through their usernames—participate in and claim a Black identity for themselves. These names also call attention to online identities that are free to claim all of these qualities and connections even as they question or revise them.

In her account for the rarity of African American scientists, Lois Powell makes the claim in an essay "Factors Associated with the Underrepresentation of African Americans in Mathematics and Science" that African American and Latino youth avoid majoring in science because their cultures have constructed negative images of scientists and believe careers in the sciences to be unrealistic goals (p. 292). Powell asserts that these unfavorable images include those of scientists as strange, unhappy, and iconoclastic, no matter how intelligent. Although Powell seems to be genuinely concerned with increasing the numbers of African Americans pursuing mathematics and science careers, the argument seems to follow the same old conservative diatribes launched against African Americans: namely, that the problem is cultural and not systemic.

BlackPlanet users respond to this perception, though, and show that Black people do connect with technologies in meaningful ways when they have access. Members "talk b(l)ack to perceptions like this by refiguring both technology and the image of the computer geek, connecting both to Black culture. Some of the forums on the site show this redefinition at work, covering topics from Linux, IT certification, tips for web designers, and Internet entrepreneurship. The language of the prompts of others might demonstrate the degree to which BP members have made technology connect with Black identity: a forum titled "what works what doesn't?" offers the following question: "What makes a good personal page? When you look at other people's personal pages on BP, what stands out?" This question might not seem significant in a discussion of cultural identity and technology, but the prompt foregrounds what other members of BP have done in establishing criteria for the broader genre of the personal web page. Another forum, "you know you are hooked on the Net when ... " begins "OK, chat heads, techies, geeks, 'puter lovers ... let's get real. Y'all know you spend way too much time on the Net when ... " Black people are techies and geeks in this forum, and their use of AAVE suggests how comfortably within Black culture. And while the joke in the prompt might suggest that Net use is discouraged, the tone and appropriation of the longstanding "you know you ghetto when" jokes show that they are accepted in the forum and on the Planet, even if they do suggest a line between being involved with technology and being hooked.

A final forum is called "Triflin Personal Pages," and leads with this comment: "Okay, I have about had it with some of these pages that are about the equivalent of Mr. T's gold chains. Too much mess!! Does anybody feel me on here?" This is a normalizing move, in which the user posting to the forum is attempting to get the group to set some expectations for page design, but of course this call looking for a response is also laced with echoes of an entire oral tradition: holla!!! Ya heard? Nahmeen?

Cross-Town Routes: Site Design and the Problem of Access

BlackPlanet's design and architecture emphasize this claiming of a separate public space, but they make that space accessible to anyone who is a part of the Planet. Although it offers a wide range of activities—almost all writing spaces of some kind, and many of them dedicated to improving technological literacy—almost everything on the site is self-contained. There are links to advertisers, of course, including gateway ads that one must click through to get to the site when one signs in, and the infamous pop up ads that have become so prevalent, but the content itself remains insular. This range of activities includes online discussions on technology, employment, relationships, gender and other identity issues; chats divided into rooms based on interests, whether "intellectual," social/sexual—"hot girls (and boys) with pics to prove it," or age-based; a personal Web page; tutorials on how to improve one's page; e-mail; news; polls; writing contests; notes and a "pager" that members use to send messages to one another. The site also allows users to search for friends (whether they are online at the time or not) based on their member profiles—a common feature on community Web sites to be sure, but one that takes on added significance with a group of people like African Americans, one with a history of rupture and displacement over generations, and often separated from other African Americans in their schools and workplaces.

Unlike many sites, the design of the homepage and the architecture of the site emphasizes the importance of all these elements. The site's structure is broad and narrow, with every Planet function either on the homepage or within one click of it.

Menu bars at the top of the page and just over the fold identify community and individual functions, respectively—chats, fora, events, channels, and games at the top; notes, e-mail, the member's personal webpage, a member find function, and the pager all at the bottom. Both menu bars are available no matter where one is in the site's structure, and remain in view even as a user scrolls below the fold of the homepage. The middle portion of the page features information more than the interactive functions presented in the menu bars, including news and polls. It also highlights content from other sections of the site, with links to the best member-written tutorials on HTML, for example, and the most popular discussion fora at the time (in this case a forum on holiday giving, taking online acquaintances offline, and the stock market).

The design of BlackPlanet is important because of what it implies about access. There are no parts of the Planet that are inaccessible because they are too far from home, or reachable only by way of endless rides on convoluted, barrier-maintaining bus systems, to use an urban planning metaphor. Nor do users have to wander around the site aimlessly in order to discover its content or get involved or connected with other Planet members, a connection that is one of the site's main goals.

HOLLA! OR, I NEED THEM HITS: COMING TO VOICE THROUGH STRUCTURED FEEDBACK

That goal, of developing an online community through various kinds of member interaction, informs the design of the interfaces of other BlackPlanet spaces as well. The personal web page and the profile that goes on it—what might be the two most important features on the entire site, in terms of the development and uses of a Black online voice—are template-driven. BP members can choose backgrounds, text sizes and colors, and receive feedback on their pages without knowing a thing about how to design, write, or code a web page. The focus here is to get people to establish themselves online as easily as possible, and then help them learn what they want to know as they go. This approach to technical communication might irk some like Johndan Johnson-Eilola, given his irritation with tacky web pages and WYSIWYG (what you see is what you get—web page editors like Netscape Composer that don't require users to know how to code their own pages) application users who are clueless about the rhetoric of web pages, as he explains in "Little Machines: Rearticulating Hypertext Writing."

But BlackPlanet's approach to teaching users would meet the exact challenge he raises as well: namely, to reestablish the social element of online writing instruction (in his case, technical instruction). Users learn from interaction with and guidance from more experienced users rather than just using a product and never learning anything about what lies behind, or underneath, the interface. This socialized instruction also contributes to the creation of a community—the exchange of feedback and guidance about personal pages helps the millions of users (10,500,000 at the time of this writing) forge connections with each other that ensure the growth of the community and the discursive conventions that develop within it.

In addition to the direct instruction that the site gives, in the way of member-written tutorials and tips on subjects like HTML, or incorporating music on one's page, or how to actually describe one's interests in interesting ways, BP members also get direct feedback on their pages from other members through the notes and guestbook entries, and indirect feedback from the hit counters on every member's page. The guestbook reinforces the metaphor that the member's personal page is her or his home on the Planet, and members regularly ask each other to visit their pages and sign each others' guestbooks. They sometimes initiate that contact by visiting random pages and leaving feedback themselves, along with a request that the recipient return the favor; or including a "tag" next to their posts in the chats asking people to visit their pages, and/or by advertising that they have pictures, tips, or other content that people want to see. Visitors to someone's "home" on the Planet can sign the guestbook, or "lick the g-spot" in the Planet's vernacular. Members crave those notes, those licks, those hits, because they provide connection with other members, because that "holla" signifies that other members were moved to respond to something that member had to say about herself or her world. In addition to comments reaffirming things a visitor liked about a particular member's page, guestbook entries and notes are used to make suggestions about how to improve the page or give technical advice on various coding tricks, or sources for graphics, links, or other content relevant to the subjects the member is interested in.

Individual member pages are so important on the Planet because they are available from everywhere else on the site. For example, when one enters a chat room, the user will see a list of the members in that room, but will also be able to link immediately to those members' pages while they chat. The discussion fora operate in the same way: they allow members to link to the page of anyone who has posted, rather than just show the username. Any search of member profiles will allow one to go directly to member pages as well, instead of providing just a profile as some sites like Yahoo! does. In other words, members' pages, their homes and identities on the Web are always available to everyone else. One can see the page of any member who has given advice on how to create a web page and see how that member incorporates her or his own advice. Members get ideas on everything from how a page should look, to the range of possibilities one has in creating a page, to how to write about one's personality and interests, and give feedback because of these mechanisms. Community works to reinforce individual identity on BlackPlanet through these functions, however similar to, or different from, members' real life identities they might be. The availability of the personal pages, at all times, a page design that incorporates multiple levels of feedback, and users who crave that feedback show that, in this online space, individual voices are still created in distinctively Black community contexts.

GOING INTO THE UNDERGROUND: AFRICAN AMERICAN DISCURSIVE FEATURES ONLINE

Knowing that there is an African American discourse community online that operates in something of an "underground" is one thing, as is to know that the design of this community, the interaction within it, and individual members' attitudes claim a distinct African American identity in the midst of theories of fragmentation and alienation are important, but that still leaves the question of the specific characteristics of the discourse that is produced in that space, and the relationship it holds with the oral tradition Smitherman (2000) describes: "as we shall see in closely examining the many facets of the oral tradition, the residue of the African world view persists, and serves to unify such seemingly disparate Black groups as preachers and poets, bluesmen and Gospel-ettes, testifiers and toast-tellers, reverends and revolutionaries. Can I get a witness?" (p. 201). Not only are all the figures Smitherman lists present in this online community, but so are all of the modes of discourse she develops in this article: call and response, mimicry, signifyin', testifyin', exaggerated language, proverbial statements, punning, spontaneity, image-making, braggadocio, indirection, and tonal semantics (pp. 217–222). The presence of these modes online show the ways that the "structural underpinnings of the oral tradition remain basically intact even as each new generation makes verbal adaptations within the tradition. Indeed the core strength of this tradition lies in its capacity to accommodate new situations and changing realities" (p. 199), but they are central to it—without these discursive practices from an African American oral tradition, BlackPlanet, a writing space, an electronic writing space at that, could not exist.

Given the fact that most attention paid to African American language and discourse in Composition has focused around AAVE or Ebonics—the grammatical, phonological, and semantic features of African American English, it is safe to assume that many interested in this subject would visit the Planet looking for these features on the pages, chats, and fora. Those visitors would find plenty of what they came for too: copula variation, distinctly African American lexical items (including, but not limited to, slang), existential it, pronunciation variation, invariant be, the absence of third person singular—s—all those features and more are present in many different places on the site. And they are often celebrated, rather than disparaged in chats and on personal pages. In fact, BlackPlanet would be an excellent linguistic site in which to study what exactly what features of AAVE do appear, with what regularity, and under what circumstances. But this isn't the issue I'm concerned with here—what's fascinating about BlackPlanet, for me, is the degree to which users have written an oral tradition into cyberspace. While I maintain that all of the modes of discourse that Smitherman identifies are present on BlackPlanet, I focus on the presence of two: tonal semantics and sermonic tone. These three features show most clearly the degree to which African American oral traditions dominate discourse on BlackPlanet.

Tonal semantics refers to the ways that intonation in a word or a phrase can change its meaning. The example Smitherman and others frequently give is of the differences that can occur in the meaning of the word police when pronounced in the typical iambic pattern of English (poLICE) and when dramatic emphasis is given to the first syllable, the *PO*lice. Part of what's involved is changes in meaning of words or phrases, and part of it is the speaker's ability to "get meaning and rhetorical mileage by triggering a familiar sound chord in the listener's ear. The words may or may not make sense; what is crucial is the rapper's [in this case, the rapper that existed before rap music: anyone can be a rapper in this sense] ability to make the words sound good. They will use rhyme, voice rhythm, repetition of key sounds and letters." (p. 222). This is a good deal to ask someone to be able to represent in writing. While BlackPlanet users certainly don't use tonal semantics to the degree that two African American speakers would in conversation, or a preacher would in addressing an audience, what is potentially surprising is the degree to which it does exist. It is used very frequently in chat rooms, with possibly as much as 30% of all utterances containing some form of tonal semantics. It is used most often in the greetings extended to people entering or leaving the room. These greetings show which members are regular members and are liked, respected, disliked, ignored, as well as new members who have not become part of that section of the Planet yet. Tonal semantics is expressed by altering spellings of words to alter tone or pitch, by shortening or elongating them to affect duration and/or pitch, and sometimes (but less often) by even appropriating the real world voices of well known figures like James Earl Jones or 1990s rapper/singer Michele' (known to have a very high, squeaky, irritating voice). Other typographic features can be used as well, like parentheses, punctuation marks—especially exclamation points, question marks, periods, and ellipses—as well as the size of the type (BlackPlanet only allows two sizes in chat, small and large). Thus, someone named "rawdawg" might be greeted by several people at once: 'sup raw/ rawwwwwwww/RAWWWW whas poppin patna????? ((((((((((((raw))))))))))))))))/ RAWWWWWWWWWWWW (((hugs))) hey baby!!!!!, where each would have different meanings and would suggest different connections between the greeter and rawdawg. Tonal semantics might also be used when one is having a semi-private conversation with a person in the public context of the chatroom, or when one is trying to get the conversation of the group.

Sermonic tone is a relatively new trope in Smitherman's (2000) canon; it refers to the ways in which plain statements are given a gravity similar to that of Sunday morning homiletic: "ordinary statements take on the tone of pronouncements and are given the force of the moral high ground; they are proclaimed with the profundity and moral sobriety of divinely inspired truth. This gives Black speech its elevated, fancy talk quality" (p. 260). The sermonic tone can be like a hyperbolic parable or fable, but without any story to illustrate its moral. The moral is taken for granted and pronounced from on high, as it were, reiterating both its simplicity and the exigence for its utterance, since obviously y'all ain't got it yet. Many times these statements are made in an elevated language register, heightening the effect, but this is not a necessary condition, as much of this tone has to do with the attitude of the speaker. The speaker has become a self-anointed preacher in these moments, having taken on the power to glorify or condemn. This connection with the church is crucial, as Smitherman points out, because it is the church that has developed—in some way—almost every African American leader in the public sphere (p. 260). Although Smitherman focuses on the figure of the preacher and points out that this leadership was largely male, this tone was available to everyone, especially outside the church.

This particular trope is very important in a discussion of African American discourse online because that identification with the Black preacher and the adoption of the sermonic tone amounts to his or her assertively claiming authority, taking permission to speak, on a subject he or she has decided has larger importance. To do this in an online space given the conditions of technology access and the ways technology and cyberspace are constructed as White, is a critical disruption of those constructions, and, again, reminders that both race and culture carry definite meaning online. And it happens all the time on BlackPlanet, both in users' own writing, and their appropriation of other texts on their behalf.

On her personal page, one user, Moonlyt1, invokes the sermonic tone in both her own words and her use of those of others:

> Be Strong.
>
> We are not here to play, dream, or drift. We`ve got hard work to do and heavy loads to lift. Shun not the struggle, for it is God's gift.
>
> Be Strong.
>
> —*unknown*
>
> *Let's chat ...*
>
> I have an appetite for more information about African Americans and our involvement in the political process. I don't believe our voice has been collectively strong enough to cut through the clutter and make a difference in our communities.
>
> Cleveland just elected its first female mayor, who begins January 7. Our Black Congresswoman is considering a bid for governor, and we may not have someone to replace her if she wins ...
>
> *What can we as African Americans do to prepare our next generation of leaders?
>
> *How do we even identify who those should be?
>
> *How do we make our voices heard to the folks currently in office?

I welcome an open dialogue on the subject. Hit me up, tell me what you think, and let me know if you`ve come across interesting web sites, magazine articles, or books I should get at.

Peace and love ...

In this excerpt, Moonlyt1 identifies herself as someone interested in political issues facing Cleveland and African-Americans in general. But through her use of the short verse "Be Strong," her descriptions of her interests, and her questions, she becomes an exhorter, attempting to foster a different kind of conversation than what normally occurs on recreational chat-based websites. Moonlyt1's voice takes on the sermonic tone, with interesting adjustments for the site's audience. Her particular voice in this space is a hybrid of formal and informal registers: the best example of this is the last paragraph: "I welcome an open dialogue on the subject. Hit me up ...."

The section begins with a general pronouncement about the state of Black involvement in politics, moves on to a specific discussion of the case in Cleveland, a city with a history of deeply-rooted involvement of its African American community and portending trouble in the future. It then moves on to questions used to make the exhortation more direct—Moonlyt1's page is not merely a pronouncement, but a call in the best of the preaching tradition. By this point the user has become the pastor issuing the conventional invitation to fellowship after the sermon: "won't you come?" Moonlyt1's invocation of the preacher's voice in her space on the Net is significant beyond the specific dialogue she hopes to spark. As important as the role of the preacher in Black culture has always been, that role has almost always been gendered as male, with women overtly prohibited from taking leadership in churches and in Black communities. Bettye Collier-Thomas (1998) provides a context for utterances like Moonlyt1's in her book *Daughters of Thunder: Black Women Preachers and Their Sermons*, noting that in spite of the obstacles to leadership in the church Black women have always "come forth to pursue the prize—the pulpit." Personal Web pages and other online writing spaces allow women to claim those pulpits and to assume the authority they seek, unfettered by the assumptions and processes that would silence that authority.

The existence of spaces where one can be comfortable with who she is and position herself as someone with something valuable to say on issues in which she has a stake—Compositionists have long made the argument that this is what matters most in developing students with strong writing, speaking, and designing voices. Although BlackPlanet is by no means an ideal site, even for African Americans, and there is much that might trouble some about the recreational uses to which people put computers and the Internet, still there is much to learn from the underground sites where people new to any discursive situation have a chance to "get their game tight:" to learn the conventions, to experiment with voice, and tone, and craft; to get feedback that they actually want from people they will listen to; while in an environment where simplistic judgments about grammatical features do not lead to their discursive and intellectual complexity being entirely dismissed and/or their continued segregation from others whose voices are often just as raw.

THE BLACKER THE DISCURSIVE BERRY, THE SWEETER THE RHETORICAL JUICE

While there is much to learn from this and other sites, the major implication of this look at BlackPlanet ought to be clear. Composition's stubborn, narrow focus on the grammatical features of a language, and insistence on waging a limiting debate on Ebonics do not work in theory or in writing practice. This stubbornness, in spite of Smitherman's (2000) own work focusing on discursive features and rhetorical traditions, and her study of nearly 3,000 student essays in the National Assessment of Educational Progress over a 20-year period showing that "the Blacker the berry, the sweeter the juice"—that the "the more discernibly African American the *discourse*, the higher the primary trait and holistic scores; the less discernibly African American the discourse, the lower the primary trait and holistic scores" (p. 184). Not just the sermonic tone, but African American discourse in general shows what can happen when students and other writers genuinely do have the right to their own language: they claim the right to speak, take the space to do it, and become invested in doing it thoroughly and effectively, and develop rhetorical savvy. As the field questions the roles new communication technologies will play in Composition theory and instruction, and begins to seriously examine questions of access, it has an opportunity to reexamine old assumptions about race, language, and technology. Neither technology, nor the English language, nor Composition *have* to continue to be White by definition.

Just as Composition needs far more study of African American discursive practices that take those discursive practices seriously, in their own right, untainted by deficit models of language use, it needs more careful consideration of how its teachers and researchers will create spaces that serve students better in a technological landscape that puts many more communicative demands on both students and teachers. To this end, this study offers three areas we can use to ask exactly how it is we want our courses to be "used" toward students' development of rhetorical and technological mastery:

1. Like the library employee, we might worry far too much about the negative results of student behavior in recreational spaces and not enough about how they might benefit from them. When we provide them with rich discursive spaces and multiple opportunities for feedback, they will often help each other come to voice, even if those spaces allow for some amount of play.
2. Related to our skittishness about student "play" is a need we still feel to control every aspect of student work and interaction. This need often results in courses that are still designed far more around surveillance and control, far more

around a willingness to do the university's police work for it, than around help students gain access to the university. We can counter this impulse toward control by including richly conceived underground spaces that students control. These spaces have encouraging student writing and design as their goals, and should give students a chance to get varied kinds of feedback they actually want from each other.

3. We still need to be primarily about the business of improving access to our courses and the university as best we can, and this is a design issue. We need to designing courses students can actually navigate, with transparent organizing schemes, interfaces that don't hide the codes that really determine how our courses work, and give students many ways to find their ways back "home," wherever they might find themselves at any given moment.

There are many other examples of potentially transformative uses African Americans have made of different technologies; these uses suggest possibilities for both making meaningful change in the designs of Rhetoric and Composition courses, and offer a hint of the kinds of knowledge users—particularly African American users—can contribute to that effort.

Rhetoric and Composition poses the exact same kind of challenge for students and faculty who take, design, and teach its courses as BlackPlanet does for its users—African Americans do not have real ownership or control in the field, even as one accounts for the tremendous accomplishments of eight former African American chairs of Conference on College Composition and Communication (CCCC). Just as African American users of BlackPlanet find ways to create underground spaces that honor and build on Black discursive practices toward larger goals of rhetorical and technological mastery, "users" of Rhetoric and Composition can focus on creating similar spaces outside the larger controls of the academy, the society, and the discursive practices that dominate the field. As this is only one exploration of one figurative planet in the digital universe, there is obviously much more work to be done before identifying specific pedagogical practices that can guide the field. But it suggests the most important work we can do on behalf of our students is not knowledge work or critical work, but design work, work in creating the spaces in which they will communicate.

This is not an argument that teachers' knowledge and composition's history of critical engagement are not important. This is not an argument that teachers can best serve students by simply creating spaces and then getting out of the way, an argument whose strongest versions are profoundly irresponsible to me. But there are times we can get out of the way and share some control, are moments when we can provide students with underground spaces, both online and off that are theirs, as one of the many goals we pursue. Stephen Doheny-Farina (2000), in his work, *The Wired Neighborhood*, quotes Tom Grundner, creator of the Cleveland Free-Net, in a vision of the Internet that can also guide Rhetoric and Composition and the spaces it designs for students:

> America's progress toward an equitable Information Age will not be measured by the number of people we can make dependent on the Internet. Rather it is the reverse. It will be measured by the number of local systems we can build, using local resources, to meet local needs. Our progress will not be measured by the number of college educated people we can bring online—but by the number of blue collar workers and farmers and families we can bring online. It will not be measured by the number of people who can access the card catalog at the University of Paris, but by the number of people who can find out what's going on at their kids' school, or get information about the latest flu bug which is going around their community. (p. 125)

Imagining Rhetoric and Composition as a kind of community network, or even "freenet" would similarly avoid becoming lost in a notion of so-called Standard English as a mythical promised land that we must deliver students to and make them buy into. Instead, the field could be guided by a vision that makes its courses dependent on their responsiveness to the wide variety of local situations students bring to them, a vision of many different paths toward the freedom that rhetorical excellence can (but does not necessarily) provide. The next chapter offers some particular principles that can guide that design for those willing to commit to it.

5

Rewriting Racist Code: The Black Jeremiad as Countertechnology in Critical Race Theory

One of the most obvious, yet most difficult sites we have to confront the technologies of race and racist exclusions would seem to be the least technological of all: our legal system. There is an inordinate number of places within that larger site begging for careful debugging, rewriting, or possibly even the destruction of the legal codes that program racism into our American "system:" constant gerrymandering and redistricting by both Democrats and Republicans to dilute Black voting blocs; three strikes laws that eliminated whatever role rehabilitation might have once played in our corrections system and turned the prison system into a wholesale market for Black and Latino labor; absurd differences in drug sentencing laws (5 years for five grams while Rush Limbaugh, one of the major proponents of those laws claimed he was ill and needed treatment when busted for trafficking illegal drugs); Proposition 209's elimination of affirmative action programs; the Newt Gingrich-led and Bill Clinton-enforced Contract "on" America with its scapegoating of the poor and people of color; the Hopwood and University of Michigan trials; Florida, 2000. To say that "the law" has been an important site of African American struggle is to risk ridicule because it's so obvious. That ridicule is worth risking, however, as the last 20 years serve as an intense reminder of how central a role the law—as legislation, jurisprudence, and the processes that make both possible—plays in maintaining racism in American society.

The examples I've cited in opening this chapter are current manifestations of just how troubled the relationships between African Americans and the legal system remain. These current examples might suggest to some that while there has been a backlash in legislation, court decisions, and legal analysis detrimental to Black progress, there is nothing systematic or racist about the American legal system. Such observers might argue that African Americans are merely facing difficult times in a system that, on the whole, works and has had neither the intent nor the effect of upholding racism or exclusions tied to race. Critical Race Theory, however, is a legal movement that has argued the exact opposite—that the American legal system, from the very beginning, encoded racism into its workings, and that the discursive conventions in both jurisprudence and legal scholarship have ensured the maintenance of that initial code. This chapter examines the ways one work in that legal movement, Derrick Bell's (1989) *And We Are Not Saved: The Elusive Quest for Racial Justice* has worked to intervene in and rewrite the racist code(s) at the root of our legal system.

My argument here is that genres and the discursive conventions that comprise them can, through their privileging of certain kinds of knowledge and experience while dismissing others, can become instruments, tools, technologies even, in maintaining established patterns of social, political, and economic relations. Form is every bit as important a site of protest as content. Bell explicitly and emphatically uses and engages the Black Jeremiad as a way of countering the instrumentality of generic conventions in American legal discourse with the purpose of forcing Black experiences and forms of knowledge into scholarly analyses of the law. Through Bell's use of the Jeremiad, he writes Black people into legal scholarship, "speaking truth to power," in ways that demand a response. Although there has been loud and ongoing praise and critique for the arguments Bell has made about fighting racism, my concern is not with the validity or the strength of the particular answers Bell's mediations on American racism offer, but on the importance of the forms those answers take.

Judge A. Leon Higganbotham (1989), in his book *Shades of Freedom: Racial Politics and the Presumptions of the American Legal Process*, calls attention to the instrumental role the law has always played in upholding American racism: "When evaluating the promises [of the Declaration of Independence, the Emancipation Proclamation, and other legal documents] 'put on hold,' I have concluded that the legal process maintained the system of slavery, using what I call the 'Ten Precepts of American Slavery Jurisprudence'" (p. xxv). For Higganbotham, despite the nation's grand ideals of equality, its practice has often been "the role of the American legal process in substantiating, perpetuating, and legitimizing the precept of inferiority," a precept that for Higganbotham meant "presume, protect, and defend the ideal of supremacy of Whites and the inferiority of Blacks. In application this precept has not remained fixed and unchanged. Nonetheless, it has persisted even to recent times, when many of the formal, overt barriers to racism have been deligitimized" (p. xxv).

Higganbotham's notion of "precepts" amount to commands programmed into a software package. The discursive conventions that carry out the work of these commands, or precepts, include the insistence that legal scholarship be "objective, neutral, and personal" and the privileged role that precedence plays in jurisprudence, which, Judge Higganbotham argues, works to maintain power relations as they are. Legal discourse, Higganbotham and Bell both show, works as an interface, a set of controls built into a system or an artifact to make it work as practically, as easily as possible while its workings remain largely invisible to those who use it. Richard and Cynthia Selfe, in their now-classic study of the Apple MacIntosh interface with its office metaphor and Beth Kolko, with her analysis of MOO/MUD environments "Erasing @race: Going White in the Interface" begin to set out the ways that interfaces can reflect and maintain racism in our technologies through the erasure of the experiences and worldviews of those who are nonWhite. For Derrick Bell, the American legal system, all the way back to its source code, all the way back to the Constitution, operates as an interface like that Apple graphical user interface: it fails to serve the needs of a huge portion of its population through assumptions about race and power relations that are hard-wired into it. His book, *And We Are Not Saved* is both a detailed examination of how racism and racialized exclusions were programmed into our legal system from the very beginning, and an attempt to intervene in it, to challenge and rewrite that code.

Derrick Bell's use of the Jeremiad (in a text that ultimately does not entirely meet the conditions of the form) is intriguing for the ways it does not meet those conditions as well as for Bell's invocation of, even dependence on the form as a means of intervening in the conventions of legal scholarship. The Black jeremiad, as a variant of the American jeremiad, is somewhat fixed in its generic conventions of citing America's and Black America's past promise through the myth of a golden age, critique of the nation's fall from that past greatness, and a prophesy that the nation must either fulfill its promise by living up to its ideals or face certain damnation. Bell manipulates and even discards some of the features of this form to make his work speak to the condition of African Americans in the legal system. First published in 1987, the bicentennial of the Constitutional Convention offers the context in which he provides both historical and contemporary critique of legal racism. In *And We Are Not Saved*, Bell confronts the rigid technology of legal discourse, which I am defining as laws, legal interpretations, and all of the discursive conventions that contribute to their creation, by acting as both the ultimate insider and ultimate outsider to those conventions. Bell is committed to using the very conventions of legal debate and scholarship while he also works to shift locus of the debate by changing its forum by introducing the jeremiad. Bell's choice to use the jeremiad in this way, to challenge legal discourse as both insider and outsider positions the jeremiad as a countertechnology. By countertechnology, I mean that Bell uses the form of the jeremiad as an instrument to unmask legal discourse, of preventing—and the human agents who employ it—from hiding behind the supposed neutrality and objectivity claimed for it in order to provoke dialogue about how to make it better serve the interests of equality for Black people.

Because language and discourses are often the main subjects of humanistic inquiry, it can seem odd to claim that they operate as technologies in the ordering of our social, economic, and political systems. But in some ways, they are our "trump" technologies, especially when specific language and discursive conventions are codified in generic conventions as they are in legal scholarship and jurisprudence. A few examples might make this connection clearer. Whether one uses MLA, APA, Chicago or any other citation system in the preparation of a document, that citation system, as a set of rules governing the appropriation and recognition of knowledge from other sources has a technological function in the academy. One's ability to use this "system" of citation often has direct effects on the kind of access she or he will obtain to academic discourse communities. The conventions surrounding Federal Reserve Chair Alan Greenspan's reports to Congress and his every informal or formal comment might be said to be technological as well: it is a set of discursive moves that are governed by specific interpretive rules and conventions that is used as a tool for ordering action in the world, or more specifically, in financial markets. Greenspan uses his comments with the knowledge that they will have such effects (whether he wants them to or not) and manages the content of those utterances as well as the timing with which he makes them and the forum in which he does so to manage various elements of those financial markets, and thus the economy at large. In these uses of language, however, there is no mistaking their instrumentality or the structuring effects they have on political, economic, and social relations in a culture.

It is still somewhat difficult to talk about discursive conventions or genres as technologies because it then becomes easy to wonder if the definition of a technology is so hopelessly muddled that it no longer defines anything. This challenge is what Charles Alan Taylor (1996), in his book *Defining Science: A Rhetoric of Demarcation* has identified as a "demarcation problem" and one that many other scholars, including Bruno Latour and Steve Woolgar, faced in their pursuits of defining scientific processes and their written products as rhetorical. Because popular understandings of technology seem to grant only specific tools—and the most current ones at that—status as technologies, it might well seem a stretch to consider the instrumentality of mere words and the codes that govern their use as technological. Computer languages, programs, and software packages help point a way through this conundrum. I have used the metaphor of codes as they work in tools like computer programs to suggest the ways that language programs action in the world; this understanding of language as having technological functions includes both the instruments and processes involved. Other language and technology scholars have confronted this definitional question, too, of course. Walter Ong (1982) has examined language itself as a technology, arguing in *Orality and Literacy* that as a system of signs it has become more technological. Jay David Bolter (1991) goes even further in collapsing writing into technology. For Bolter, the different "spaces" in which writing occurs (papyrus rolls, codexes, print, computers) are each technologies, as the process of writing in each of these spaces entails different sets of techniques for negotiating relationships between writers and readers.

One notable move toward clarifying this muddle comes from Langdon Winner, in his 1986 book *The Whale and the Reactor.* Winner argues that technologies are not simply the instruments, the tools that we touch, use, save, or throw away. He discussions technologies as both "objects and processes" and argues that "the crucial weakness of the conventional idea is that it disregards the many ways in which technologies provide structure for human activity." Winner adds that "technologies are not merely aids to human activity, but also powerful forces acting to reshape that activity and its meaning" (p. 6). I take Winner to mean that the computer—and what we do with it, what it comes to mean in the society—is both aid and powerful force, but that structuring forces that, while instrumental, do not make their way into boxes and wires are also tools.

If technologies can be processes as well as artifacts, and if they can be more than mere tools because they have important effects on the structures of the societies that use them, then it is not such a stretch to consider a genre and its discursive conventions to have technological functions—especially one that is so clearly both process and artifact, tool and shaper of society, as legal discourse is. Judge A. Leon Higganbotham made a career long scholarly project of documenting these aspects of legal discourse in the maintenance of slavery in his histories of legal process *In the Matter of Color* (1980) and *Shades of Freedom.* In the quote from Higganbotham cited above, Higganbotham sets out to examine the "pernicious role" that the American legal process has played in maintaining racial inequality in the United States. In spite of the assumptions many hold that the law is objective, and that racism is simply the result of individuals who are ignorant and must be persuaded to be more "tolerant," more "diverse," more "multicultural," Higganbotham works to show that racism and racial exclusions are indeed systemic, nurtured, and supported by technologies like the legal system in ways that allow individuals to deny responsibility for its existence, because after all, every American now claims wholeheartedly to believe in "equality." The "precepts" that Higganbotham identifies as the linchpins of the American legal system that allowed it to maintain slavery and the racism that followed are significant because they reflect the one element of any technology that is crucial to its utility, but is almost never discussed: the understandings about social, political, and economic relationships that guide how that technology is used. Higganbotham, then, offers two definitions of precept, that connect it with the computer programming metaphor at work in this chapter: "a command or principle intended as a general rule of action" (p. 3), and more directly related to the legal process, "a rule of law, a legal principle, and a legal doctrine" (p. 4).

Higganbotham uses this concept of a precept to show the ways that racism against African Americans and their exclusion from almost all facets of American life were "hard wired" into the American legal system and have not been dismantled, even to this day. Bolter helps make the comparison to computer programming more explicit: "programming is embodied logic: the establishment of logical relationships among symbols that are embodied in and empowered by memory chips and processors of the digital computer" (p. 9). In the case of the legal system, Higganbotham helps to show how racism is the logic programmed into laws and legal interpretations—the ways Black people become symbols for inferiority and guilt, the ways legal precedent becomes the "memory chip," and judges and juries the "processors." In introducing the precept with which *Shades of Freedom* is concerned—that of the inferiority of Black people—Higganbotham shows how these assumptions of Black inferiority (assumptions whose roots I take up in chapter two) and guilt have become so deeply rooted in the code of law as to be hidden. These hidden operations call up Jeffrey Walker's (1994) definition of the enthymeme in his article "The Body of Persuasion: A Theory of Enthymeme." Enthymeme, for Walker, is "a complex structure of intuitive reference and affect that constitutes the substance of an argument; and ... a structural/stylistic turn that caps an exestasis, gives the inferential/affective substance a particular realization with a particular salience for a particular discursive moment, and by doing so constructs or shapes its audience's perception of just what 'the argument' is" (p. 48).

The language gets a bit complex there, but we all know what he means: "states rights" allows some right-wing democrats and republicans since Strom Thurmond's Dixiecrat campaign for President to signify a wide range of arguments for policies detrimental to Black people without the indelicacy of being forced to refer to nigras or niggers or Negroes or "you people." Attacks on crime during the 1970s, 80s, and 90s get used to pass and implement laws that continue to disproportionately imprison African American and Latino men. Martin Luther King Jr.'s language appealing to a nation to finally get beyond its fixation on race comes to stand for legions of arguments for us to be "colorblind" now, and therefore remove even the basic protections Black people and other people of color were provided as a result of Civil Rights and Black Power struggle. Higganbotham uses the workings of the Supreme Court in the introduction to his book to describe the role of jurisprudence and the effects of his precepts as enthymematic bits of code that work to ensure the enforcement or racism:

> In official judicial opinions, Justices of the Supreme Court generally use a careful protocol, which masks their inner feelings about what they know to be the egregiously unfair treatment of African Americans ... [Shades of Freedom] traces what Chief Justice Burger described as the "agonizingly slow" process of the Civil Rights movement. It also looks at, to again use Chief Justice Burger's telling phrase, the "pernicious role" that the Supreme Court and other Whites in power played when the promises of the Declaration of Independence, the Emancipation Proclamation, and the Civil War amendments were put on hold. (1996, p. xxv)

The importance of Higganbotham's work to African American rhetoric is that those who struggle for changes in law or policy on behalf of those who are still denied educations, jailed unfairly and disproportionately, killed by police, denied employment opportunities, or prevented from voting do not face a simple task of "persuading" those in power to make the changes for which they lobby. In addition to the overt persuasion of interested individuals and groups, there is also the job of dismantling the tools, or the parts of those tools that allow injustices to occur. Although Judge Higganbotham carefully describes the workings of legal process as programming racism into American life, Critical Race Theorists continue that work by attempting to dismantle and rewrite those codes by explicitly attacking the conventions of legal discourse.

Both Critical Race Theory and Critical Legal Studies have set out to debunk the assumptions that "the law" is objective, and that justice is simply a scientific exercise in which one can plug in a fact pattern (without subjective considerations of how the actual parties stand before the law) before any judge and get just the same result. For literary scholars, this might sound eerily similar to strong versions of formalism and the New Criticism of Cleanth Brooks, W. K. Wimsatt, and Monroe Beardsley. The overall point of this school of legal interpretation is that the law has, and should have, no role in addressing systematic inequities, or in providing any kind of redress to the victims of such inequities. The current version of that thinking is embodied in President George W. Bush's 2000 campaign as well as his 2004 reelection campaign pledging only to appoint "strict constructionists" to the federal bench. "Strict constructionist jurists," in addition to arguing that subjective considerations are irrelevant to the interpretation of laws, have often voted clearly and consistently against the interests of people of color, reproductive rights, and citizens with disabilities. This particular approach to jurisprudence, a far-reaching version of legal formalism, is best represented in figures like Charles Pickering, Antonin Scalia, and Clarence Thomas. Often jurists of this ilk reject this kind of labeling, but the phrase has clearly become, again, enthymematic, working to cue a set of arguments about policy and power relations without actually having to make them. Critical Race Theorists, like Critical Legal Studies proponents might be described as legal realists: scholars and jurists who argue that the legal system must be responsive to those who have had no voice in the legal system. They argue that the law must be more proactive to address injustices in American society.

Critical Race Theory grew out of the Critical Legal Studies movement, when, according to Richard Delgado and Jean Stefanic, legal scholars began to see that lawsuits and legislation were being used to systematically roll back the advances of the Civil Rights and Black Power movements. The group held its first conference in 1989, but has been active since the mid-1970s (Delgado & Stefanic, 2000, p. 4). One of the most distinguishing features of Critical Race Theory has been its challenge of legal formalism on a formal level, on a discursive level. Critical Race Theory attacks the conventions of legal scholarship, even while rigorously practicing that scholarship. Legal writing—especially scholarly legal writing—is supposed to be absolutely impersonal, entirely objective, thoroughly neutral. Critical Race Theorists like Derrick Bell and Patricia Williams took dead aim at the rigidity of these conventions by boldly infusing the personal into their analysis. The effects of a supposedly neutral and objective legal process on those who have been consistently marginalized became relevant, as did the passions and anger of those who were asked to stand idly by, because, of course, the legal process is fair. Patricia Williams made it even more clearly personal than did Bell, injecting the "Diary of a Mad Law Professor," as she titled her column on legal issues for *The Nation*, insisting on her individual agency, and the collective agency of African Americans. Other scholars, like Girardeau Spann, Gerald Torres, Angela Harris, Regina Austin, and more, have followed suit (and laid down precedents) in ways equally jarring to a staid, comfortable legal community.

Derrick Bell's first book, and the book that some mark as the beginning of the mainstream popularity of Critical Race Theory, goes further by using parable, allegory, even science fiction in this project of dismantling the systemic supports for legal racism. Despite its mainstream popularity, however, *And We Are Not Saved* is intriguing exactly because it is not written entirely, or even primarily, for a mainstream audience. Bell's book, a series of ten "chronicles"—Socratic dialogues between Bell and a mythical character, Geneva Crenshaw, on just how difficult the goal of racial justice is to achieve, is laid down as a challenge to those least willing to take it up: a legal community strenuously opposed to the strategies he employs and the goal of racial justice he employs them toward. The chronicles are arranged to provide an analysis across the sweep of American legal history, showing how thoroughly racism determines major aspects of life for members of all racial groups—and how thoroughly the legal system is implicated in that history. Thus, Bell connects elements of racism and its history that seem to have nothing to do with the legal system to issues still facing African Americans, in a legal and public climate where those issues were frequently dismissed.

Those strategies and the book that resulted have demanded a response, forced dialogue, and therein created at least the possibility of a genuine rhetorical situation, where none existed before. This is the value of Bell's book and the work of Critical Race Theorists in African American rhetorical study—not the ultimate judgment of whether one thinks Bell's book actually solves the problems of racism or systematic differences in access to the technologies that make up American society. Detine Bowers' (1996) article "When Outsiders Encounter Insiders in Speaking" describes the way Black rhetors often have to create a rhetorical situation from nothing: the "task is an ongoing attempt to argue for our right to equality and our right to speak before getting on with our particular policy claim—that is, when we can make the policy claim. Such claims are usually left untackled because matters of collective definition take precedence. The predetermined condition for speech, one that accepts voices of the oppressed only on the definitional terms of the power structure" (p. 490). One particular example of this phenomenon occurs in the ways African American interests are often collapsed under the heading of "special interests," as if questions of Black participation in the economic and political processes of the nation are concerns on the same level as gun lobbyists or environmentalists. The examples of Martin Luther King and Malcolm X from chapter three present this phenomenon a bit more clearly in the ways they were often required to defend themselves against charges that they were communist or anti-White *before* having any opportunity to present their appeals for the right of African Americans to vote or work.

The central premise of *And We Are Not Saved* is that rather than a perversion or temporary irregularity that is the result of ignorant people who only need somehow to be enlightened, racism and the marginalization of African Americans are systemic and a crucial part of American history. Bell argues through legal analysis that the exclusion of Black people from the processes that make up the nation were the supports that made American nationalism, and literally, the nation possible. The chronicles, imagined conversations between Bell and Crenshaw represent the best of legal scholarship, complete with rigorous citation of laws, court cases, and case histories, presented in narrative form. In the book's fictional storyline, in the middle of a civil rights organization's annual convention, Crenshaw "visits" Bell, contacting her old friend and asking him to help her decide on the future of Black struggle. Crenshaw reaches out to Bell, or his character, because she has been appointed to a kind of judgment day panel that will once and for all determine the path that struggle will take. Crenshaw's panel is stuck trying to decide between engaging in armed struggle or separating from America, because it has become obvious that the long history of American racism and racial exclusion will not change without drastic action. These questions, of course, have always been the central questions that African Americans have debated, from Phillis Wheatley and Prince Hall to the slave-era convention movement, to Martin Delany and Frederick Douglass, to T. Thomas Fortune and Frances E. W. Harper and Edward Wilmot Blyden, to DuBois and Washington and Marcus to Mary McLeod Bethune and Claudia Jones and Paul Robeson, to King and Malcolm, to even today. Crenshaw and Bell arrive at a third solution: to continue on, in nonviolent struggle, and here in America, but with the purpose of not only winning justice for Black people, but of transforming the nation. Although there are myriad differences between Derrick Bell on one hand and Manning Marable on the other, with one of the most glaring being that Marable's analysis of race is deeply rooted in the history of class struggle, there are many similarities between Bell's answer and that which Marable and Mullings attribute to those engaged in labor and class struggle: "this perspective accepted neither the structure of the contemporary society nor called for separate Black society, but rather advocated a radical transformation of the United States based on a fundamental redistribution of resources" (p. xx).

Aside from the significance of Bell's use of the jeremiad toward these ends, his book should be central to studies of Black rhetorical traditions for several reasons. First, his decision to take legal arguments and make them in a more public discourse, the decision to open up for public scrutiny discursive practices that amount to a perpetual closed door, "executive session" of a legal system makes clear the limits of that system without resignation to it. Bell, in fact, becomes the archetypical lawyer in this book, taking the obscurantism upon which legal arguments often depend and "making them plain," in the best of the "Country Preacher" tradition that led Cannonball Adderley and Joe Zawinul to mythologize Jesse Jackson in the early 1970s. This ability, to articulate in clear terms the legal theory on which a movement was created, to open up that theory and the uses to which it is put for examination and dialogue by those who have been excluded. This is exactly what we need more of, not just for Blackfolk, but throughout the entire society. Imagine if someone were able to bring this combination of clarity and advocacy to the policy decisions made by the Federal Communications Commission throughout its history, or to the possibilities Open Source might hold as a theoretical basis for African American struggle and its connections with other forms of anti-racist, anti-capitalist struggle, as well as ensuring that individual technology users and school systems are not trapped into allegiances to one platform, one software package, one system. In addition to this two-edged sword of attacking the foundations of structural racism encoded into our legal system and making that attack accessible to many outside the legal community, Bell lays down this challenge to the legal system by invoking one of the major forms of the African American rhetorical tradition—the Black jeremiad. While *Saved* is not in itself a jeremiad, Bell's invocation of the form, his use and engagement of it as a rhetorical strategy in this situation, intervenes in and counters the generic rigidity of legal scholarship. It not only tells a legal community to "step outside," to a figurative place where its inner workings might be more clearly seen, but calls that community's members onto Black rhetorical ground, where African American discursive traditions are foregrounded, Black ways of knowing encoded.

The Black jeremiad, according to Wilson Moses' (1978) study *The Golden Age of Black Nationalism, 1850–1920* and to David Howard Pitney's extension of Moses' arguments in his book *The Afro-American Jeremiad: Appeals for Justice in America*, is an African American rhetorical form that is both a warning and a lament, named for the biblical prophet Jeremiah. The Black jeremiad is a variant of the American jeremiad, which breaks down into three parts: "citing the promise, criticism of present declension, or retrogression from the promise; and a resolving prophesy that society will surely complete its mission and redeem the promise" (p. 8) or face damnation. The African American variation of the form is marked by its incessant use, the way it became a "ubiquitous rhetorical convention," with which African Americans represented themselves as a chosen people. In this case, African Americans became a chosen people among a chosen people, and argued through this form that America's sense of itself as a chosen people was only as valid as its ability to deal fairly with Black people. The Black jeremiad as a form was linked with a "messianic Black nationalism." The irony of such close connections between radical Black nationalisms and White American versions of nationalism is not lost on Howard Pitney, who notes that Black reliance on the form "signals their virtually complete acceptance of and incorporation into the national cultural norm of millennial faith in America's promise" (p. 13). The Black jeremiad was not stripped of meaning as a tool of resistance, either, however, as it reiterated a Black destiny distinct from that of the larger nation, showed faith in that destiny, and served as a means for Black people to assert both their separate (and separatist) ambitions while still yearning for meaningful inclusion in the larger nation (p. 13).

Despite the jeremiad's importance to Black rhetorical traditions, the use of the form alone in this situation in itself is not what makes it noteworthy. In fact, Bell's use of it in 1987 would seem to evoke the same mixed review Howard Pitney offers as mentioned, as 1987 was certainly a high moment in American nationalism as well, with a political landscape dominated by notions of the United States as "the greatest nation in the history of the world" in a cold war with the "evil empire" of the Soviet Union. Again, the use of this form in such a nation whose identity was (and is) still dependent on the construction of African Americans and other people of color as somehow existing outside the legal process, unworthy of its protections, is just a bit ironic. The fascinating element in this book is Bell's clear use of it while he wrestles with the form's implications. Bell partially invokes the Black jeremiad and its history while he cannot bring himself to argue that America had a glorious past to fall from. Nor can he commit to the warning that accompanies the form of damnation should the society fail to heed the jeremiah's warnings, so his use is at once more radical and more conservative than what is already a radical and conservative form. Bell relies on both the optimism created by the belief in a mythical golden age and the pessimism of the threat of apocalypse, though he cannot bring himself to state either directly. Thus, the specter of real Black jeremiahs is called up—jeremiahs like David Walker, who surreptitiously sewed copies of his *Appeal* into the clothes of his customers while pronouncing not only were African slaves in America "the most degraded, wretched, and abject set of beings that ever lived since the world began" (p. 23) and catalogues that brutality while arguing that the nation should be ashamed of allowing such conditions given its ideals, but that "they must look sharp or this very thing will bring swift destruction" (p. 33). Bell invokes Black jeremiahs throughout America's history to deliver both the carrot of a past worth reclaiming and the stick of certain damnation as he remains noncommittal to both, attempting to cajole the legal community and the larger nation toward redemption.

More significant than the genre-blurring use of allegory and science fiction in Bell's near-jeremiad is its complete divorce from the conventional appeals to America's promise: the noble ideals of its founding and faith in its ultimate victory over these and all other problems that face it. The first Chronicle disintegrates the nation's founding myths of a birth in noble, universal, "self-evident," even, truths that will withstand all competing political forces into a tale of a nation delivered into a hell of continual compromise by "forefathers" completely aware of the compromises they made. In this first chronicle, "The Chronicle of the Constitutional Convention," the heroine of the book, Geneva Crenshaw, is transported through time to the Constitutional Convention of 1787. Her task is a version of a thought experiment played out many times among civil rights lawyers and activists: "sup-pose we could recruit a battalion of the best Black lawyers from the era of Brown v. Board of Education—Thurgood Marshall, Robert L. Carter, Constance Baker Motley, Robert Ming, William Hastie, Spottswood Robinson, and Charles Houston—and send them as a delegation back through time to reason with the Framers before they decided to incorporate slavery into the Constitution. Surely that impressive body would influence the decisions made by those who knew Blacks only as slaves?" (p. 24). The job then, for Crenshaw, is the ultimate test of the limits and/or possibilities of rhetoric: if the delegates to the Convention were provided with foresight about the results of decisions they would make, *and* evidence embodied in Crenshaw herself refuting the assumptions of Black inferiority developed in the Enlightenment (assumptions addressed earlier, in chapter two) and built into the legal system, *and* impeccable arguments against the "pragmatic" reasons for such a compromise, could those delegates be persuaded to create a founding document, and therefore a nation whose creation did not depend on the systematic exploitation of Black people?

No. This imagined conversation between Crenshaw and delegates to the Constitutional Convention takes predictable turns. Completely unwilling to hear her—her presence as both a woman and a Black person being just too much to bear—the delegates demand that she be ejected, a demand that would have been fulfilled were it not for an electric force field that surrounded her to protect from that very likelihood. The Framers challenge her authority, asking how dare she insinuate herself into conversations that do not concern her, and then, gently, "sympathetically," inform her that while they appreciate the difficulties she has raised, they have already thought them through quite carefully and reasonably, and come to the best possible solution given the situation. The situation is exactly as Bowers describes—even after the Framers believe that Crenshaw has come from the future, and after they are forced to acknowledge her ability, her equality, she must still justify her presence and argue for the right to speak at all before being patronizingly dismissed from the conversation.

Despite the fact that the Framers clearly ain't havin' it, clearly are not trying to hear her arguments, and despite the fact that she is only protected from violent removal by her force field, Crenshaw presses on in her attempt to help them see the dangers the nation would face as a result of their impending actions. The delegates respond to her arguments as Bell imagines they would, but he does not leave their responses to his whims. He bases them instead on the actual words of some of the delegates to the real convention in 1787, and his historical analysis of the general arguments made at the time. Despite the range of responses to Crenshaw, from mild, sometimes condescending sympathy from some northern delegates to the outright scorn of those from slaveholding states, and the complete silence of some like George Washington, the result remains the same: the reduction of humans to the level of property and the shrink-wrapping of that reduction into the user-agreement for the new nation. The protections of those property relations still influence the status of Black people in American society, because every attempt to repair them is caught up in continual contest and resistance. When Crenshaw is finally heard, the answer remains the same. Bell summarizes the responses in the historical record in the words of a fictional, unnamed delegate:

> You have, by now, heard enough to realize that we have not lightly reached the compromises on slavery you so deplore. Perhaps we, with the responsibility of forming a radically new government in perilous times, see more clearly than is possible for you in hindsight that the unavoidable cost of our labors will be the need to accept and live with what you call a contradiction. This contradiction is not lost on us. Surely we know, even though we are at pains not to mention it, that we have sacrificed the rights of some in the belief that this involuntary forfeiture is necessary to secure the rights for others in a society espousing, as its basic principle, the liberty of all ... But what alternative do we have? Unless we frame here a constitution that can first gain our signatures and then win ratification by the states we shall soon have no nation. For better or worse, slavery has been the backbone of our economy, the source of much of our wealth. (pp. 36, 37)

The response Bell crafts for this fictional delegate amounts to a synthesis of the debates that took place in many different spaces by real Convention delegates and others. The point of the synthesis and the historical analysis that allows Bell to offer it is that the Framers felt this compromise to be necessary, regardless of their intentions in making it. The charge that the nation was founded on the exploitation of Black people to ensure liberties for some is a far cry from the narrative of God-given, unalienable rights guaranteed to all and vigorously defended by America's birth and rise to power. This refusal to accept as a given the myth of America as a nation founded in great ideals and chosen to bring justice (usually in the form of "democracy") and safeguard it here and abroad is why *And We Are Not Saved* is possibly the more important of his two departures from the jeremiadic form, again, even as he depends on those resonances and uses them creatively.

The significance of this departure is difficult to overstate, as Howard-Pitney explains that the implicit acceptance of the myth of America as elect "on one hand placed important limits on these groups' resistance and efforts at autonomy; but on the other hand, these cultural norms provided an ideological shield behind which women and slaves steadily advanced their rights and freedom ... it also illustrates the shrewd and artful tendency of an oppressed group to refashion values taught by privileged classes—even as it accepts them—into ideological tools for its own ends" (p. 186). By directly challenging America's myth of an origin located in a golden age, Bell abandons the possibility of pointing to a glorious past that can be recaptured as motivation for the changes he seeks, and risks not being heard in the first place because of his audience's possible rejection of his claim of a nation born in compromise rather than nobility—especially in a moment like the celebration of the 200th anniversary of the Constitutional Convention and the patriotic sentiment that surrounded it.

This compromise, for Bell, not only marks the founding of the American nation, but is the defining feature of responses throughout out its history to questions of the legacies of slavery and racism, when mainstream America has responded at all. Watershed moments like the Civil War, Reconstruction, and the Civil Rights Act are the subjects of the second chronicle, "The Chronicle of the Celestial Curia." In this Chronicle, Crenshaw is asked to become the third member of a tribunal that will decide the future of African American struggles for justice and equality. She has been invited to help mediate the strenuous disagreement between the other two members, one of whom advocates armed struggle against Whites while the other presses for emigration to another country because it has become obvious to the Curia (and to many involved in civil rights activism during the 1980s) that traditional strategies, including the litigation that resulted in many of the most impressive civil-rights era victories, could not ensure justice or equality for African Americans.

The particular plan they ask her advice on is one called the "Conservative Crusader," a scheme designed to incite unrest among both people of color and White communities by placing a staunchly conservative justice at the head of the Supreme Court, a justice who proceeds to attack all of the most basic protections provided for those who are marginalized. The ensuing anger, if the plan works, would be enough to finally call the question of which of the Curia's paths African Americans should take—that of armed struggle or mass exodus. The reasons for considering this plan are several. The current state of declension that marks the second major trait in the jeremiad has been constant, even with the advances of the civil rights era. Those legislative and judicial victories, so widely recognized as landmark moments where America finally got it right, in their half-hearted implementation, amount to "social programs, which even now manage only to stave off starvation while keeping the masses too weak to recognize their true status" (p. 54) and "protecting upper-class privilege ... to offer the poor food without nutrition, welfare without well-being, job training without employment opportunities, and minimal legal services without real expectations of justice" (p. 55). Bell also points out in this Chronicle the oft-repeated observation that insurrection and defiance produce far more than commitments to working within the system of electoral politics for those who have been excluded from the system (p. 58).

In the end, Bell and Crenshaw dismiss the plan, because, as has been made amply clear during the years following the book's publication by a William Rehnquist-led Supreme Court and a Black conservative member that some hoped would fulfill the very function of the Curia's "crusader," it is not only difficult to spark and sustain large-scale protest, but that the courts themselves can only play a "limited" role in addressing the problems Bell and Crenshaw hope to help solve. The analysis that follows in the Chronicle shows that not only have the courts been a conservative force in American politics, but that they have only made reforms on issues of race when they also presented a clear benefit to Whites and/or cost them little—that those reforms required no sacrifice in the name of racial justice.

The dialogue in the Chronicle offers as support the analysis of Lewis Steel, a NAACP attorney during the 1960s and law professor Arthur Miller, which together reveal the deeply rooted conservatism that protected White interests from more radical remedies in decisions like Brown v. Board, but also the reluctance of African American leadership to offer any systematic critique challenging these patterns in the Supreme Court's decisions. The point for Bell here is that, even the "one place of redress" available to Black people, the one space where they could even make the argument for equality could only offer answers that "invariable promote the interests of the White majority" (p. 63). Far more reflection on case histories and precedents follows on issues like voter disenfranchisement, due process, and the equal protection clause of the Fourteenth Amendment, demonstrating the ways Black struggle results in the courts serving White interests—even when addressing distinctly African American grievances. Litigation as a strategy, then, is at best an "endless detour" for African Americans (p. 69), offering only temporary victories that do not primarily benefit Black people, and are always easily and systematically eroded. The compromise that founded the nation continues throughout its legal history, replacing the notion of a fall from lofty heights with a despair that even the imagined myth of a nation that values equality cannot be rescued.

Other Chronicles engage this legal history in far more detail, looking at the restraints preventing equal access to the franchise, public education, and economic opportunity, as well as the effects of this legacy on Black lives. The last chapter of the book, "Salvation for All: The Ultimate Civil Rights Strategy," contains "The Chronicle of the Black Crime Cure." This Chronicle is meant to remove all doubt about the depth and breadth of the challenges that face those assembled in the book's fictional civil rights conference, to serve as a final warning as they contemplate the most effective strategy for the future of their struggle. While other Chronicles address topics like affirmative action, the equal protection clause, Black male and female relationships, and the social status of Black people, as well as many potential strategies for addressing these collected problems, "The Chronicle of the Black Crime Cure" is meant for those who think that Black pathology rather than systematic problems resulting from slavery and racism is the source of the persistence of African Americans' problems in the United States. In the Chronicle, a cure is discovered for Black crime, and the cure is total: "the converted drug dealer himself not only immediately lost all inclination to wrongdoing but was possessed with an overpowering desire to fight Black crime wherever it existed ... Black people were overjoyed and looked forward to life without fear in even the poorest neighborhoods. Whites also lost their fear of muggings, burglary, and rape" (pp. 245, 246).

But even this does not have the effect some might expect: the end of Black crime changed neither public policy nor private behavior toward Black people. In fact, the role of the construction of African Americans as disproportionately involved in crime becomes more clear: "by causing Whites with otherwise conflicting political economic and political interests to suspect all Blacks as potential attackers, the threat persuades many Whites that they must unite against this common danger. The phenomenon is not new. In the pre-Civil War era, slave owners used the threat of violent slave revolts as the common danger to gain support for slavery among Whites, including those who opposed the institution on moral grounds and those in the working class whose economic interests were harmed by the existence of slave labor" (p. 247).

Bell's point, for many, is obvious: no matter how distressing the situation of African Americans in the prison system, no matter how much crime rates seem to go up or down, the construction of Black people as inherently more criminal than other groups of people remains, even to the present. Drug laws and their enforcement are only the most extreme site of this hypocrisy. Manning Marable, in *The Great Wells of Democracy: The Meaning of Race in American Life* notes the results of such constructions. African Americans account for only 14% illegal drug users, but 55% of all drug convictions; African Americans and Latinos combine for 25% of the nation's population, but more than sixty percent of prison populations, begging the question "are Blacks just unlucky in the courts, or is something else at work here?" (p. 4). Angela Davis examines the relationships between these inordinately high conviction and prison rates of people of color, poor people, and immigrants, the privatization of prisons, and the erosion of the few programs that helped to comprise the "safety net" that once existed to protect the poor. With populations eight times levels of only three decades ago (p. 54), the prison has become the new plantation. Davis quotes Eve Goldberg and Linda Evans as saying

> Prison labor is like a pot of gold. No strikes. No union organizing. No health benefits, unemployment insurance, or workers' compensation to pay. No language barriers, as in foreign countries. New leviathan prisons are being built with thousands of eerie acres of factories inside the walls. Prisoners do data entry for Chevron, make telephone reservations for TWA, raise hogs, shovel manure, make circuit boards, limousines, waterbeds, and lingerie for Victoria's Secret—all at a fraction of the cost of 'free labor.' (p. 57)

Regardless of whether one wants to accept the connections Davis makes, the compromises inherent in forcing Black labor and dehumanizing Black lives continue.

The final Chronicle presents the other departure from Moses' and Howard-Pitney's model of the Black jeremiad, the belief in the nation's ultimate salvation—its ability to return to the glory of its golden age. While this point is already moot on some level, with no golden age available to motivate belief in a glorious future, the belief in some possibility is important, as optimism, no matter how muted is necessary to compel future action. The mood among the delegates to the bicentennial of the Constitutional Convention was understandably dark at this point—the Chronicle "touched the very nadir of our despair. It stimulated no discussion. Rather, a pall of resignation fell over the gathering" (p. 248). Obviously, this tension requires some resolution, and Geneva asks the Celestial Curia why she or anyone should be subjected to the Chronicles "if their only message is that our civil rights programs are worthless opiates offering no more than delusions of hope to a people whose color has foredoomed them to lives of tokenism, subservience, and exclusion" (p. 249). The Curia's answer to Geneva's challenge is what keeps Bell's departure from the conventions of the jeremiad from becoming a complete disavowal of the form.

The charge for those concerned with justice and equality for African Americans and other marginalized groups is the same relative conservatism that led Malcolm X and many others to despair of traditional leadership in the movement. The guidance from the Curia amounts to imploring Black people to keep on keepin on: looking "away from your earthbound yearning for equal opportunity and acceptance" (p. 250). In other words, African Americans should continue the struggles they have engaged in, but with the goal of transforming America for everyone else rather than hold any hope of sharing equally in its benefits, because to seek an equal share of opportunity and resources given the way the nation is currently constructed is to become complicit in the exploitation of others. The Curia argue that just as African Americans transformed the Bible from a text that was used to maintain slavery and racism to a foundation of a faith based on freedom, they should do the same, attempt the unthinkable with the Constitution and the legal system. Thus, the responsibility for the nation's salvation depends on African Americans, as it always has in the Black jeremiad, but without the reassurance that comes with identification as a chosen people, because the notion of America as elect has been discarded. Nor is their success assured, as no prophecy of ultimate victory ensues. Only the possibility of success exists, and the self-assurance of commitment to greater principles, in spite of the daunting and self-sacrificing task ahead (pp. 253–255).

The conclusions that are the results of Bell's reflection on America's past and possibility are troubling, in spite of the incisive analysis that leads to them, as they contain both the best and the worst of traditional Civil Rights era thought. Black struggle, as shown in the continual use of the jeremiad, has always been about the transformation of the entire nation, and Bell continues in this tradition. The final argument, however, that African Americans must forget their thoughts of equal access and the redress of longstanding injustices is exactly the otherworldly perspective that wore thin with many elements of the Black community during the 1960s. David Howard-Pitney (1990) addresses this tension by pointing out the paradox in the form itself. It is "both radical and conservative" (p. 186). Although Bell tries mightily to dispense with the more conservative elements of the form, he cannot. Let me be clear here: the attempt to transform a nation by dramatically reinterpreting its history and its future, can be both liberating and radical. And to be sure, as Martin Luther King noted in his famous "I Have a Dream" speech, unrecognized work on behalf of others, like unearned suffering, is redemptive. But to ask the Black community to continue to be America's John Coffey, the gargantuan hyperbole of a character in Steven King's *The Green Mile*, subject to all of Americas racism and social ills, deprived of freedom, and still asked to suck the venom out of those social ills, with only the hope of some internalized redemption, is to ask it to be the ultimate codependent. Bell is roundly criticized for these conclusions, however, and he includes some of those conclusions as discussion starters in the paperback version of the book.

Ultimately, however, Bell's attempt at the transformative is what makes his use of the jeremiad valuable, not so much because of the troubled conclusion of the book, but because of what his use of the form, of form more generally, as a battleground for countering the effects of racist assumptions about African Americans codified into the documents and tools and processes of the American legal system. David Howard-Pitney concludes his book on the jeremiad by responding to Wilson Moses' claim that growing Black frustration with America's racial intransigence will lead to a decline in the form and a lost of the faith in the nation that is its signature characteristic. Howard-Pitney offers an analysis of the rhetoric of Jesse Jackson to show that the form and its optimism will continue to be integral to rhetorical manifestations of African American struggle:

> There is nothing new about America's failure to deal forthrightly with the challenges of racism and social justice or about the betrayal of frequently soaring Afro-American patriotic expectations. High hope is a necessary condition for grievous disappointment. Afro-American hope for America has often run low but eventually has welled up to burst forth again. Scholars have erred in interpreting the phenomenon of Black disappointment in America as a one time event of the 1920s, late 1960s, or whenever; it seems more accurate to regard the ebb and flow of Black hopes for America as a recurrent historical pattern—at least until now. If the past is the best guide to the future and the Jackson phenomenon indicative at all of contemporary Black attitudes, then it is premature to forecast the demise of the Black American jeremiad and the resilient hope on which it rests. (p. 194)

Derrick Bell's *And We Are Not Saved: The Elusive Quest for Racial Justice* attempts to cling desperately to "the faith that the dark past [and present] has taught us" and yet refuse its requirement that jeremiahs accept the ideological givens of the society he or she hopes to change. This struggle, only three years after Jackson's famous "Keep Hope Alive" speech at the 1984 Democratic National Convention, is a mightily different use of the form than David Howard-Pitney demonstrates in Jackson, and could show that both Moses and Howard-Pitney are right. Bell's use of the form might reveal a long-term break with the abiding faith that Moses argues is the legacy of the 1960s *and* a dogged, enduring commitment that makes it too soon to claim that the form is on the decline.

The future of the jeremiad aside, however, it is clear that a major challenge in fighting racism and pursuing access for African Americans depends on the ability of those fighting that fight to root out and attack the coded assumptions about Black people that allow that racism to continue to work, about dislodging the precepts that are placed so deeply into our societal structures as to seem invisible. Form is every bit as important a site of protest as content. In fact, appealing again to Andrew Feenberg (1995a), this time in *Alternative Modernity: The Technical Turn in Philosophy and Social Theory*, technological systems are not objective facts that just happen to always work out in favor of those who already have power at all: "technological design is socially relative, contrary to deterministic arguments or theories of technical neutrality; the unequal distribution of social influence over technological design contributes to social injustice; and there are at least some instances in which public involvement in the design of devices and systems has made a difference" (p. 3). Derrick Bell's use of form to attack and expose the instrumentality of encoded racism in the American legal system calls us, as Feenberg does, to pay more careful attention to codes and processes and their designs. The next chapter looks for the map Bell leads us to, but ultimately cannot provide.

6

Through This Hell Into Freedom: Black Architects, Slave Quilters and an African American Rhetoric of Design

But how do we get from systemic critique to making some difference in the design of technological systems? This question points to a significant blindspot in African American rhetorical study as it's currently conceived. It is imperative that such study address design issues, however. This need is so pressing not only because design itself is rhetorical, but because the history of design has been so exclusive and has done so much to enforce the very exclusions encoded in the nation and its technologies. Design needs to be as important a site of struggle as schools, ballot boxes, and police practices. Greg Tate and Arthur Jafa begin to explore these questions, and the relationships between visual traditions and language in an article "From Dogon to Digital: Design Force 2000. Looting Other Disciplines Along the Way," arguing that Black scholars and artists need to be as thorough in documenting visual and design traditions as they have the spoken and written word:

> We know the continuity is there. You can trace it directly in music; you can trace it directly in language. When you start talking about visual culture it gets a little difficult. A little difficult? Or is it a way of seeing? Perceiving the artful, expressive nature of forms conventionally (in Western culture) held to be unsightly or absurd: hair styled to stand straight up in appreciation of kinks' sculptural defiance of gravity? Seeing precise, zigzag parts in a girl's hair as more interesting than straight parts? Loving asymmetry, contrariness and exaggeration: just one pants leg rolled up; caps worn backwards, at a slant, or squashed in a back pocket; taut physiques dribbling down the courts in dragging, triple X shorts; the odd juxtaposition which is ultimately, in its own way, right on. Are there "cool," "clean" (the perfect

persona projected with ease) and "funk" formulations that are applicable to the design process?

Or is it a way of seeing? Arthur Jafa and Greg Tate (1998) ask the perfectly timed question. How we make a difference in the design of technological systems is, at least partially, a question of how we "see" our work. For Jafa and Tate, we need to see design differently by looking at our own traditions to make that difference, to interrupt those practices and theoretical approaches that continue to exclude African Americans and other people of color. Amiri Baraka (1972) implored people to see these same connections, more than 30 years ago, explicitly grounding technology design in both Black culture and rhetoric in a short essay called "Technology and Ethos." The essay deserves to be quoted in its entirety, but the opening sentences issue the call even more dramatically than Jafa and Tate:

> MACHINES (as Norbert Wiener said) are an extension of their inventor-creators. That is not simple once you think. Machines, the entire technology of the West, is just that, the technology of the West. Nothing *has* to look or function the way it does. The Western man's freedom, unscientifically got at the expense of the rest of the world's people, has allowed him to expand his mind—spread his sensibility wherever it cd go, & so *shaped* the world, & its powerful artifact-engines. Political power is *also* the power to create—not only what you will, but o be freed to go where ever you can go (mentally physically as well). Black creation—creation powered by the Black ethos brings very special results. (p. 319)

Baraka asks probing questions about how African American communicators might harness the potentials of technologies in work and play, how Blackfolk would address health crises, even what the "Black purposes of space travel" might be. The point is that Baraka connects Black technological mastery not only with distinctively Black exigencies, but with Black traditions as well, deadly serious and seriously playful as he makes the connections—as only Baraka could:

> So that a telephone is one culture's solution to the problem of sending words through space. It is political power that has allowed this technology to emerge, & seems the sole direction for the result desired. A typewriter?—why shd it only make use of the fingers as contact points of flowing multidirectional creativity. If I invented a word placing machine, an "expression-scriber" if you will, then I would have a kind of instrument into which I could step & sit or sprawl or hang & use not only my fingers to make words express feelings but elbows, feet, head, behind, and all the sounds I wanted, screams, grunts, taps, itches, I'd have magnetically recorded, at the same time & translated into word—or perhaps even the final xpressed thought/feeling wd not be merely word or sheet, but *itself*, the xpression, three dimensional—able to be touched, or tasted, or felt, or entered, or heard, or carried like a speaking singing constantly communicating charm. *A typewriter is corny*!! (p. 320)

In his own articulation of a Black technological ethos, Baraka insists that individual and collective self knowledge are crucial, and that African American pursuits of technological innovation must proceed from that point of self knowledge, and completely destroy our current notions of form in order to see again, to be able to imagine and produce new forms, free of the assumptions and exclusions embedded, encoded into old forms and old technologies. It is, in short, the same Black digital ethos presented in chapter three in connection with Martin Luther King and Malcolm X: rooted in African American traditions, committed to African American freedom, and willing to completely reimagine everything that our society is, can be, and should be.

Despite the rigorously exclusive nature of design professions and their active role in enforcing and maintaining racism, and in spite of the possibilities that Baraka shows design offers in the pursuit of freedom, Black design traditions that challenge and counter these roles have, as Jafa and Tate note, been difficult to see, difficult to chart. So difficult, in fact, that architect Melvin Mitchell (2001) argues in his book *The Crisis of the African American Architect: Conflicting Cultures of Architecture and (Black) Power*, that Black architects—perhaps the most visible of the minuscule presence of Black design professionals—have yet to articulate a specifically African American approach to design in the attempt to solve these problems. In this chapter, I examine design professions' ugly history (and present) in upholding and enforcing racism and begin to chart some of the traditions of Black designers to demonstrate how design fits squarely within African American rhetoric and can be an important part of the move from critique to design that must become a part of African American struggle in this new millennium. I argue that a synthesis of the work of African American architects and the slave quilts used as a major element in the Underground Railroad offer a distinctly African American rhetoric of design that both preserves Black visual traditions and offers possibilities for transformation. As contrasted with notions of a specific visual or design aesthetic, by African American rhetoric of design I mean a set of principles for design taken from the work of people who used their work to aid some sense of collective struggle, a kind of heuristic that can be applied to design processes regardless of the different aesthetic choices one might make. This rhetoric of design is bidirectional, working on both sides of transformation, and would challenge designers to build freedom and pathways to it into technologies and spaces. In other words, this approach to design would challenge its practitioners to, like the slave quilters, design the signs and sign systems to lead people to freedom into the artifacts themselves, and like Black architects attempted to do, design and build spaces and technologies that take freedom for granted, creating spaces in which Black people, once free, can live, work, play, worship, and communicate as free people, no matter how far off a genuine freedom might seem to be.

One of the reasons it's so easy to critique the conclusion of Derrick Bell's *And We Are Not Saved* and ask questions like why other Black jeremiahs have not effected more change for Blackfolk is that the jeremiad as a form is essentially limited to critique. Jeremiahs are the rhetorical demolition experts challenged to create space for dialogue, for a genuine rhetorical situation where none existed before. Given such a charge, it is difficult to ask those same individuals to then step into the space they've cleared and create new spaces, new technologies, even a new nation, as well. People's desire to get both the implosion and the blueprint from the same source at the same time, though, is intense. So intense that it often becomes palpable at public events where speakers purport to address the "race problem," as it was often called in the 19th and early 20th centuries. At these events, even now, regardless of whether the invited speaker has national, local, or no renown, the very first question or comment—and the theme of much of the ensuing discussion—is some kind of variant of the question "so what do we do now?"

As Wilson Moses and David Howard-Pitney show in their work on the jeremiad, it is something of a conservative form (even in its most radical instantiations) because it is a form of public engagement. This is the case not only because jeremiahs often operate in the panopticon of government, media, corporate interests, communities, and individuals who have always opposed a society that genuinely attempts to solve the problems of justice and equal participation for African Americans. Jeremiahs are also forced to issue that cry in the wilderness that must be heard, and are thus forced to bear the responsibility for creating dialogue in a society that often senses no need for such dialogue—having reduced appeals for racial justice to "special interests," or metaphorically diminished Black struggle to a game, a "race card" that Black people supposedly "play" in order to obtain advantages unfair to the rest of the nation.

The jeremiad and other forms of public critique are arguments for access, for that proverbial place at the table that Malcolm X both wanted and refused to be satisfied with. The work of design is different altogether. It is not public work—at least it should not be entirely public for African Americans. It is not for those who represent the public faces of Black struggle, whether they be Malcolm, Angela Davis, Al Sharpton, David Walker, or Ida B. Wells. Those public figures, at least in their public roles of demanding that the codes and interfaces of America and its technologies be changed and made just for all its citizens, cannot be asked to then go write the laws, design the freeways and buildings, plan new urban communities, and transform the computer industry. Critique can clearly be transformative, but the actual work of planning and designing transformed spaces and technologies has to be underground work, even when public figures participate in it. It has to be underground work (or at least partially underground) because those who do it must be able to do so free of what bell hooks has often called the "gaze" that is always focused on Black people. It must also be as free as possible of the constant contest and exclusion that mark discussions of race. This is the case even as I acknowledge that the actual process of building new spaces, new technologies, new nations, is a public process that will of necessity involve both context and compromise. It is also underground work because those who do it, in many ways, have been forced underground by virtue of their exclusion from design professions. Of the various design professions, only "urban planning" has any semblance of an African American presence, and that presence is often muted by the networks of power in which they are forced to participate. Other design professions, like architecture, are 97% White. Organizational and industrial design statistics are as bad or worse. Further, design work is underground because it is often done before and outside of more public phases of the production of any product or space.

Of course there are different perspectives on design which oppose this underground approach. The most notable of these is participatory design, which is a user centered approach that attempts to remove the veil that separates design from production and consumption. This approach is, at its best, an attempt for a genuinely democratic and inclusive design, and one that I advocate. The design of computer products, community spaces, buildings, highways, and laws needs to be opened up. At the same time, however, groups of people who have been excluded from these processes for as long as African Americans have need to nurture spaces away from the public gaze in order to develop the interests they would then bring to democratic, user-centered design processes. In fact, the closed, exclusive, system-centered approaches to design that so dominate American culture are in large part responsible for the exclusions that mark our society and its technologies now. Langdon Winner (1986) provides an example of the real effects of this kind of exclusion as the result of the combination of racism and undemocratic design processes, in his recounting of the story of New York City planner Robert Moses. Winner describes the way Moses, by design, explicitly limited that access people of color and poor people would have to his widely acclaimed "public" park, Jones Beach. He not only designed his overpasses at such a height that New York's twelve-foot tall buses could not handle them, but also vetoed a proposal to extend the Long Island Railroad to Jones Beach (p. 23). Beyond an individual example, however, the rhetorical nature of design, its ugly history, and the need for African Americans to take it up as a site of struggle with transformative possibilities all come together in William Mitchell's 1995 manifesto *City of Bits*. The extended reading of Mitchell's well-received work would seem to some to be both harsh and long, especially considering that Mitchell has made scholarly statements on issues of technology access by co-editing the collection *High Technology and Low Income Communities: Prospects for the Positive Use of Advanced Information Technology* (1999). The point of such an extended reading, however, is to show in detail the assumptions that lie beneath the exclusions to design professions and processes that African Americans face, even from those who purport to have good intentions.

STRANGERS AT THE GATE: A RHETORIC OF DESIGN(ED) EXCLUSION IN THE CITY OF BITS

First published in 1995, the same year that the Digital Divide entered public discourse, *City of Bits* deserves a careful reading not just because it does an excellent job of selling the wonders of cyberspace without engaging the exclusions that it both maintains and creates, but also because of the role that utopian rhetorics play in selling these wonders. Lisa Nakamura, in her essay "Where Do You Want to Go Today? Cybernetic Tourism, the Internet, and Transnationality" (not on reference page) looks at this kind of rhetoric in television commercials and other discursive spaces. Nakamura argues "this world without limits is represented by vivid and often sublime images of displayed ethnic and racial difference in order to bracket them off as exotic and irremediably other. Images of this other as primitive, anachronistic, and picturesque decorate the landscape of these ads" (2000, p. 17).

It might be easy to dismiss television ads as the work of evil, money-grubbing corporations who really don't care about those who have been denied access to the technologies they sell. But the same thinking is at the core of Mitchell's book, although he would see himself as someone who does care about problems of access. Just as people who did not own property were not considered citizens in the Greek society with which Mitchell is so enamored (just as are many of us who study and profess Rhetoric are still), those without technology, without bandwidth, are left as strangers at the gate—entirely different and utterly irrelevant.

REACHING INTO THE GOODIE BAG: AN OVERVIEW OF THE CITY OF BITS

Ladies and gentlemen, daddy's home. The war is over, and Odysseus has arrived, bearing gifts for all of the citizens of the new Greece. This is what Mitchells's *City of Bits: Space, Place, and the Infobahn* amounts to. The theory that Mitchell presents in the book, the city of bits hypothesis, pronounces that current technologies will foster sea changes in the way we live at least as great as those fostered by other technological marvels: the railroad, aviation, the telephone, and even the printing press.

In his book, Mitchell announces that he not only has new toys for us to play with, but a new world to play in as well—a world of ease based on the changes that the Internet will bring us. This virtual world hinges on the metaphor of the city, and virtual societies must be planned as seriously as the brick and mortar ones with which we are so familiar. For Mitchell, those of us who would take on the challenge of planning the new utopia (one of his other books is titled *E-topia*) have much to learn from architects and urban planners. The benefits Mitchell believes will result from this world depend on the emergence of new kinds of programmers who will design spaces in much the same way as urban planners and architects do, in order to turn the frontier of the Internet as it existed in 1995 into the new kind of utopic *polis* he envisions.

We should know by now that there is no such thing as utopia, except in the stories that we create. And Mitchell's story is just as dangerous as it is compelling. It actively excludes people, and although not racist, serves to maintain the kinds of racial inequities in access that Selfe and Moran describe, and needs to be challenged as such. Whether the result of a neglect born of naivete, or an ideological cyst that must be operated on to have any hope of saving what might otherwise be a healthy body, his theory and the book in which he presents it suffers the same failings as the urban planning theory on which he depends. In his vision, Mitchell is all too willing to continue to ignore those who have always been ignored—poor, non-White, rural, and even urban people, in the worst of ironies, given the importance of the city as metaphor to that vision. Problems of access, when mentioned at all, are glibly pointed to in asides, often literally marginalized by parentheses. At least urban planning professionals sometimes deluded themselves into thinking they were addressing these problems directly, that they were attempting to serve the people they spent much of the 20th-century planning into poverty and exclusion. And even the parenthetical comments and asides Mitchell does offer are laden with deficit models of social theory, further attacking their victims.

Still, it's hard not to get caught up in the overwhelming promises Mitchell makes for those of us willing to buy into his theory. He offers a world of convenience for all in which physical location no longer matters, and human cognition and existence are no longer embodied. But, like the infamous Ginsu steak knife commercials, Wait! There's more! We can create our identities any way we want, no longer subject to the exclusions of our current world, and we will have our very own little postmodern helots to follow our every command. If we really do know that there is no such thing as utopia, what makes Mitchell's city so attractive and so believable? It's a simple answer, actually. We like big dreams, even when we know we're dreaming, and Mitchell gives us something to dream about. At the very least, Mitchell's city looks good before one can look at it more carefully.

What changes attractive to compelling and makes Mitchell's city seem probable if not certain is a rhetorical strategy that highlights his own investment in the argument, as an architect, as an academic, and as someone who cares about technology. Rather than play the ad writer, changing the product and pitch with each new poll or market niche, Mitchell sells his theory consistently, passionately, shrewdly. The savvy in this book is the result of two connected tactics: a sharp ability to lean on both recent and ancient history; and an ability to join huge, sweeping panoramas of democracy realized with an all-encompassing web of the minutiae of convenience. To beat up on a cliché, we will have our Nikes and be able to wear them too—without having to worry about who made them, how we got them, or who went hungry so we could have them.

There really should be a temporary moratorium on references to Greece as the epitome of democracy realized, especially in conversations about community, the city, government, or the larger society. Maybe a year or two will calm things down. That's how hard Mitchell leans on his metaphor of the city of bits as the perfected metropolis: "the network is the urban site before us, an invitation to design and construct the City of Bits (capital of the twenty-first century), just as, so long ago, a narrow peninsula beside the Meander became the place for Miletos ... Who shall be our Hippodamos?" (1995, p. 24). In his recruitment speech for city planners to design the spaces that will shape communities on the Internet, Mitchell proclaims that the Net will be "the agora, the forum, the piazza, the café, the bar, the bathhouse, the college dining hall, the common room, the office, or the club" (p. 7), as well as our new work and home spaces.

Far more the ideal community than Greece ever could offer, but Greece remains the model from beginning to end: "it will play a crucial role in twenty-first century urbanity as the centrally located, spatially bounded, architecturally celebrated Agora did (according to Aristotle's *Politics*, 1956) in the life of the Greek polis and in prototypical urban diagrams like that so lucidly traced out by the Milesians on their Ionian rock" (p. 8). When we finally "get to the good bits" of Mitchell's last chapters, we hear more about civitas, urbs, how "the ancient polis provided an agora" like what we will soon see in cyberspace, and questions "worthy of an online Aristotle" (p. 161). What is so troubling about the role that Greece and the ancient polis play in Mitchell's argument is not just that it romanticizes the past. We know the Greece that the Western academy loves and adores is gone. The connection is powerful—in spite of the fact that we know ancient Greece was far from an ideal society—because the argument is that we can do even better than our most romanticized versions of the Greek ideal. We don't have to worry about messy border wars or sloppy ethics questions about our use of slaves. Contest over spaces and bodies simply disappears with our click of the "connect" button.

These sweeping claims about new communities and a new democracy alone would not mean much, however, because we've heard all of that before in one way or another. But Mitchell also offers close readings of the technologies we already have, contextualizing inventions like the Walkman, television, telephones, and hearing aids, to both ground the development of the Internet in a recent history of modern technology over the last century and a half, and to extend the narrative into the near future to tell us what we will soon have.

We will not only have technology to do both important and menial tasks for us, but we will have technology to manage our technology. Mitchell describes in great detail the "bots" or computer programs that already handle some functions like sorting e-mail and posting messages to newsgroups and suggests that they will soon negotiate and make purchases for us (like Ebay auction trackers, for example, but more intricate and involved), continually growing in "intelligence" and capacity to do for us what we don't feel like doing for ourselves. But, still, there's more. Mitchell also offers visions of "resort offices" and corporate headquarters on mountains as possibilities that will emerge from the growing importance of bandwidth and the diminishing role of physical location. The troubles of the outsourcing of jobs and the systematic underdevelopment of urban areas that plague us now will soon mean nothing because soon we'll be able to move entire businesses and industries wherever we might want. The post World War II and 1970s/1980s phases of White flight into inner, and then outer ring suburbs, gated communities, and restrictive subdivisions were nothing compared to what we'll now be able to do!

Living spaces will be transformed as well, with "intelligent appliances" that "radiate information instead of heat" (p. 99). But the goodie bag runs even deeper: even the human body itself will change because of wearable and implantable technology connected to the Internet—data gloves, electronic jogging shoes, Dictaphones, and "anything else that you might habitually wear or occasionally carry—can seamlessly be linked in a wireless bodynet that allows the objects to function as an integrated system and connects them to a worldwide digital network" (p. 29). All fun stuff, much of it even potentially very helpful and hard to turn down, as the excitement over cyborgs in much writing about technology tells us. But what's the cost of buying into the vision?

WHAT'S WRONG WITH DREAMING BIG? THE CITY OF BITS' DARK UNDERBELLY

Mitchell's utopianism and the toys that construct it might be easier to take if they were simply a part of some fairy tale by and for those who might not have known better. But the book is not merely a game of pretend or a thought experiment. There are several problems with the vision Mitchell presents, not the least of which is the extremely passive view of the Internet and technology that informs the theory that comprises it. Worse, the imagined city that has Mitchell so rapt fosters, directly and indirectly, the same exclusions with which we still struggle, with not even an attempt to employ technologies in the search for solutions to them.

Despite all we know about the failures of the "city"—all of the poverty that still exists, all the crime that results from the desperation people face, all the separation that our approaches to urban planning have fostered, and in spite of the fact that race is still a major determining factor in all of those problems, Mitchell is extremely comfortable grounding his vision of the Internet in this metaphor, with no attempt to question either the limits of the metaphor or ways design professions are implicated in maintaining poverty or racialized exclusions. More important, those who are currently marginalized in our society remain that way in Mitchell's: all of the improvements that these new technologies will bring, and the wonder associated with them, will serve only those who are already in the network, those who already have access to technologies. Mitchell's electronic window from MIT to Oxford, Xerox PARC researchers, Bill Gates' home, and CNN's uses of technologies are simply much too interesting for him to carefully consider the implications of his frame of reference, or to even wonder aloud how networked technologies can help to solve fundamental problems like racism, poverty, violence, and unequal patterns of technology access.

To give credit where it is due, Mitchell does put his vision to work on some problems. For example, law enforcement becomes easier for the state (and supposedly the criminal) in Mitchell's cityworld: "electronics can now perform many of a prison's traditional functions without cells and walls—discipline and punishment sans slammer" (p. 78). Lest one think that Mitchell is starting to present an argument for more humane law enforcement, however, he also adds

> Of course the system would not be complete without effective ways to apply immobilizing force and punitive violence. But that doesn't seem

> too difficult. Anklets could automatically sound loud alarms when triggered by entry to forbidden places or when activated by wardens. There might be some behavior-monitoring capacity built into an anklet or an implant, together with a drug-release mechanism; one advocate of walking prisons imagines that "a sex offender's specific patterns of aberrant sexuality would be recognized by the programmed chip, and the drugs would selectively tone down criminally sanctioned behaviors but allow normal or acceptable sexuality." For maximum-security offenders, the drugs could be sleep-inducing or lethal. (p. 78)

It doesn't matter to Mitchell that his facile comments on the state's "legal monopoly on confinement and violence" (p. 78) describe a system that consistently, systematically, targets people of color, and especially men of color. He has no interest in this matter—in how technologies might exacerbate that racism—nor even in how they might help to eliminate it. The technologies he describes might just as easily release drugs into the systems of violent and malicious police officers so that they become more likely to use the minimum force necessary in subduing suspects rather than the maximum allowable force after the adrenaline-pumping experiences of car or foot chases, to ensure at least some degree of humanity for those suspects. Of course, that is not Mitchell's vision, nor would the forced drugging of violent police be accepted as uncritically as the continued abuse of prisoners.

Ruminations like these throughout the book and the absence of any reflection on how technologies might help to undo problems like systematic racism or end poverty and police brutality is troubling enough. Mitchell at times makes casual, often parenthetic comments about the importance of access as an issue in the creation of the Infobahn without engaging that issue in search of solutions, in ways that smacks of elitist navel-gazing (but with the good intentions of at least mentioning access, of course). However, Mitchell quickly and repeatedly disabuses readers of any notion that this exclusion is the result of a mere ivory tower or geek culture ignorance. The contempt for those who have been excluded that often lies just beneath the surface of public discourses about education, law, government, and technology rises to the surface throughout *City of Bits* to let those strangers at the gate know exactly why they are being ignored.

To be sure, Mitchell has brief moments when he allows thoughts about social equity to infect his prophecy:

> shall we allow home-based employment, education, entertainment, and other opportunities and services be channeled to some households and not to others, thereby technologically creating and maintaining a new kind of privilege? Or can we use the Infobahn as an equalizing mechanism—a device for providing enhanced access to these benefits for the geographically isolated, the homebound elderly, the sick and disabled, and those who cannot afford wheels? (p. 103)

But even in this rare moment, which serves more as a comfort to those who are already included in the vision than to those who are not, the book maintains exclusions in identifying those who are worthy of assistance and those who are not. In this case, only those who are disabled or physically isolated are worth the energy of thought; there is still no role for technology in addressing old, ingrained patterns of privilege.

And these moments fade. This small gesture, whatever its worth, is taken back rather quickly: "if we can no longer make the traditional urban distinctions between, on the one hand, major public and commercial buildings, and, on the other hand, relatively uniform and repetitive housing areas, how shall we make social organizations and power legible?" (p. 103). This passage shows the real power, and the real problem in Mitchell's metaphor, and his project. Instead of an image chosen for its gleaming downtowns even as it ignores the poor who languish unseen and unheard in them, the city as metaphor is important to Mitchell's project precisely because architects and urban planners can make sure that the Infobahn bears the markings of power as it is reified in our physical spaces.

Of course, Mitchell would likely respond by arguing that the issue here is maintaining the elements of culture and society that any group of people needs, that "making social organizations and power legible" is much less threatening than it sounds. The sentence previously quoted, however still seems to make the argument that the crucial task of the Internet is maintaining traditional social and power relations between people. This is especially so when taken in light of a statement just a few pages before: "such instabilities and ambiguities in space use also challenge traditional ways of representing social distinctions and stages of socialization … At an urban scale, prisons, convents, residential colleges, orphanages, halfway houses, official residences for politicians and religious leaders, and low income housing projects make vivid social distinctions by creating readily identifiable, physically discreet domains" (p. 103). Mitchell might maintain that he is neither right nor left politically, and that he cares about social equity. While the city is his controlling metaphor, cyberspace challenges it and will at some point transform it (p. 107). However, readers are again disabused of such generous readings "perhaps it is not too romantic to imagine that unique natural environments, culturally resonant urban settings, and local communities that hold special social meaning will increasingly reassert their power" (p. 104). The project of settling the cyberspace frontier and creating the Infobahn is about maintaining distinction, power, and current patterns of social relations.

So the city isn't really new at all. This point is made particularly clear in a reflection on the benefits technology will have for the health care industry. Remotely controlled robotic devices are good for surgical tasks that require great precision, of course. But then again, such devices are also good because "you might just want to stay well away from dangerous battlefields of the South Side of Chicago" (p. 38). The snide tone of the remark is offensive enough, but the comparison of an urban area to the dangers and inconveniences of active minefields, bomb detonation sites, or the "infectious samples in a laboratory" assures us that the poor and people of color who have been urbanly planned and architecturally designed into these lives and areas will be untouchables in the new cyber world as well.

ARCHITECTURE, URBAN DEVELOPMENT, AND THE PRESERVATION OF "LEGIBLE POWER"

Regardless of any analysis one might offer of any specific text, we all know we live in an environment in which no one will admit to actively contributing to the silencing and removal of marginalized groups of people. So no one would expect Mitchell to admit as much either. To be fair to him, he would counter that he does at least raise issues of access, and that these issues are far too difficult for any one person to solve in one book (although he attempts to solve nearly every other problem of contemporary life in the broad sweep of his vision). And he might be right. Maybe we can't blame an architect for upholding centuries of racism and the failings of urban policy. It might even be that Robert McLeod was right in his introduction to Harry Siedler's (1978) *Planning and Building Down Under: New Settlement Strategy and Architectural Practice in Australia.* Australia, as Mitchell would know, and as many others know, has had monumental problems with the confluence of race, geography, and architecture throughout the 20th century. In that introduction, McLeod asks very important questions about the actual role architects and designers play in upholding racism and other exclusions versus the role we might imagine them to play:

> One of the great burdens that active and influential architects *must always bear* is the incorrect and unreasonable identification with all of the overwhelming forces that shape buildings before pencil is ever put to paper. Did they create the consumptive and corrosive "commodity market" in urban land that forms our cities? Did they invent and perform the irrational systems and modes of transport and services that choke and besmirch our settlements? Did they create the adversary relations between those who commission buildings and those who use them, between those who inspect and those who execute? (pp. ix, x, emphasis added)

It seems, as McLeod believed when this was written—in 1978—that the political, social, and economic forces that structure racism and exclusion are simply too interconnected and too strong to single out any one text or any one person for criticism. But even if one agrees with him that the architect is not responsible for the ways they make those patterns "legible" in our buildings and public spaces, none of the allowances he makes apply to a world that Mitchell partly perceives and at least half creates. Mitchell's role in this book is not that of an architect designing individual buildings and spaces at the behest of some patron. He is beholden to no politicians or commissions in this vision. He doesn't have flawed transportation systems about which to worry (unless you count the bandwidth problem, which he does not do sufficiently). And Mitchell does not have to deal with the problems of free market systems and the ways they can make access even more prohibitive a barrier to the spaces he foresees, as he ignores those forces and their effects.

Through his use of the city as metaphor for his version of cyberspace and the new virtual and physical worlds it will usher into existence, Mitchell becomes architect, planner, developer, zoning commission, and politician, with all of their combined power—at least for the length of the fantasy that is his book. By selling us a world, he is at least partially responsible for what that world contains. The limits and problems of Mitchell's vision cannot be easily dismissed as his alone, however. As an architect and urban cyber planner (not to mention technoculture insider), he is, of course, heir to very steeped traditions or racists who used the tools of their trades to create power and wealth, maintain it for some, and deprive it from many. Yet these are the fields he argues are crucial—models, even—for the successful development of the Internet and the societies it will create and change.

Gwendolyn Wright (2000) opened a graduate course in the history and theory of architecture at Columbia University by telling her students on the course syllabus page that "urbanistic practice, both public and private, encouraged racial and ethnic segregation, functional zoning, and sizable investments in infrastructure and industrial areas" (http://www.arch.columbia.edu/admin/syllabi/a4529.html). An April 12, 1995 *Yale Daily News* story reports on a symposium of urban planners discussing racism in architecture, asking if people "ever wondered why janitor's closets are located in the back and not the front of buildings? Or why dumpsters are located on alleys and not streetfronts?" The city planners who spoke at the symposium included John Jeffries of the New School for Social Research, who noted the seeming permanence of racism, and indicted design professions in that permanence, adding that "the only way the separation of the races is legitimated is by the designation of places within the city as racially separate." Harvard sponsored an exhibit and conference in 1998 titled "Architecture and Segregation" that toured U.S. cities in 2002 to systematically explore these issues as well.

The point here is that there is and has been work done to show the relationship between architecture, urban planning, and the lived experiences of people—not just by those who have been systematically excluded, but from individuals on the other side of the drafting board as well. Other writers have studied racism in urban policy: George Galster and Edward Hill (1992) in their edited collection *The Metropolis in Black and White*, Howard Gilette (1995), in *Between Justice and Beauty: Race, Planning, and the Failure of Policy in Washington, D.C.*, and William Goldsmith and Edward Blakely's (1992) *Separate Societies: Poverty and Inequality in US Cities*. Lesley Lokko's (2000) collection *White Papers, Black Marks: Architecture, Race, and Culture* looks specifically at architecture in maintaining racialized exclusions. These books, courses, and conferences, reveal disciplines that have been aware of these relationships, and have begun to take responsibility for the effects they have had on the American landscape.

Mitchell is not guilty of perpetuating race and class-based exclusions in his book merely because of his place in a profession that has carried them out. His controlling metaphor of the city, however, and his argument that virtual architects and urban planners should do for cyberspace what they have done with and to cities, with no willingness to critique or even qualify either the metaphor or the practices of those in his profession while heralding the Internet and digital technologies as the end of all our problems is both exclusionary, and disingenuous, regardless of the author's intentions.

Baraka reminds us why all of this is so important to African American rhetoric, as well as to rhetoric, composition, technical communication, and Black studies: critique alone will not interrupt these practices. Those of us who care about ending systematic oppressions must design new spaces, even as we point out problems in our current ones. While there is probably a need for people like the virtual architects Mitchell evokes to design spaces and communities on the Internet, there is an even greater need for people to make sure that the development of cyberspace doesn't simply replicate the kinds of systematic exclusions that make up the unseen, rarely interrogated infrastructure of our uses of physical space. The problem *is* access, and any vision of what cyberspace can, will, or should become that does not seriously engage this problem cannot make good on any other claims it might make, no matter how attractive. Scholars in composition and technical communication have begun to see access as an important intellectual problem, and some have even begun to see it as a rhetorical problem in addition to a material one. Some of those scholars have also begun to pay serious attention to design in their teaching and theorizing, as the Internet has made the visual far more important than it ever had been to whatever we now call "writing." Our ability to ensure that problems of access and design come together to help people move from no access, to the kinds of passive, consumer and corporate driven access that Mitchell and others understand Internet users as needing, to a more meaningful power that holds the potential to transform experiences, lives, spaces, and technologies depends on far more than the ability to write for, speak for, advocate for, those who have been locked out of opportunity and planned out of community. It depends on the ability to get people involved in answering design and policy questions for themselves and understanding the networks of power that prevent designers, policy makers, politicians, educators, and others from making more equitable design and policy decisions.

DESIGN AS RHETORICAL?

Although there are those on the computers and writing and technical communication sides of rhetoric and composition who have begun to take up questions of design and access as important subjects of study, the field still grapples with just how and to what extent they belong. The emergence of computes and internet related technologies as dominant communication tools has sparked several debates about what the goals of composition should be, and a seemingly endless list of new terms for those possible goals: electronic literacies, web literacy, information literacies, visual rhetoric, rhetorics of design, and even essayistic literacy. Of all these terms and their associated claims the goals and purposes of writing instruction to emerge within the last decade, one that might pose the most challenge to the way writing instruction (and by extension, rhetoric and composition as a field) is conceived is that of design. Despite the seeming incongruence with what we're used to doing that design might pose, African American rhetoric, as well as the field more broadly, should embrace design, even as one of its major concerns. A focus on both traditional literacies and design can help move rhetoric and composition beyond some of the tired debates that have consumed it throughout its history, and toward a more unified sense of its mission with areas like technical and professional communication. Not only is design rhetorical, but it can help fill in some of the spaces literacy simply cannot in courses whose aims include helping students to engage in some kind of productive action in the world.

The debate itself, around what place, if any, design should have in composition, is still in its early phases. Some of this debate has taken place without explicitly mentioning design at all, but rather through a general set of questions about what composition should do for students in a more highly technologized age—questions that frequently focus on literacies. Among the more interesting treatments of these questions of what roles literacies and design should play are Michael Joyce's (1995) book *Of Two Minds: Hypertext Pedagogy and Poetics*, as it is one of very few that actually take up the issue of how writing instructors might actually integrate design into their classroom practices. Myka Vielstemmig's (1999) "Petals On a Wet, Black Bough" deals with design questions from a print paradigm, pushing the limits of what one can do with design in print, academic articles. Anne Wysocki and Johndan Johnson Eilola's (1999) "Blinded By the Letter: Why Are We Using Literacy as a Metaphor for Everything Else?" and Doug Hesse's (1999) "Saving a Place for Essayistic Literacy," both appearing in Passion, Pedagogies, and 21st Century Technologies, offer a debate on the relationship between literacies and design, between print paradigms and digital, multimedia ones worth engaging.

Wysocki and Johnson-Eilola's article takes on the task of clearing room for discussions about design as a major element of writing instruction by polemicizing the place of "literacy" as the main trope used to talk about the goals of writing instruction. They argue that the term literacy has become so overused as to become cliché—that it is the first term anyone grabs onto when attempting to talk about any kind of knowledge used toward any goals in any situation. After outlining the ways they feel literacy has been stripped of meaning from this overuse, Wysocki and Johnson-Eilola then argue that for writing teachers to continue to focus on literacy (no matter what those teachers think it means) is not the empowering move they tend to think it is, but rather, is limiting because of the ways it operates as a myth of guaranteed success. Clearly this is an overly aggressive read on the intentions of those who use literacy as an important trope in their work. Literacy as a tool of empowerment, even as a tool necessary to empowerment has always been a huge element of African American struggle, but the severity of racism would prevent anyone from that tradition from dangling it as a guarantee of any kind of success in the United States. This tradition, from the trope of the talking book that operates as an originating mythos for African American literature, to its deployment in artists like Phillis Wheatley, Frederick Douglass, and legions more, has always focused on the acquisition of literacy as one part of a larger individual and collective journey to empowerment. Literacies, as Elaine Richardson demonstrates in her book on the subject, have always been, and will continue to be central to individual African American empowerment and collective Black struggle. Wysocki and Johnson-Eilola's point, however, that the term is used so much by so many in so many contexts, with little clarity about what it means, is well taken. And as they show in their essay, there are so many other things we can do in conjunction with literacy to make writing instruction more valuable.

Doug Hesse understands these concerns about the vagueness and overuse of concepts like literacy, but wants to hold on to both "literacy" and the essay as the major foci of composition. Hesse attempts to define the essay and argue for its relevance in a technology-driven curriculum by showing ways that digital genres like the web page and listservs can be essayistic. Ultimately, the essay and "essayistic literacy" appeal to Hesse as a needed check on the immediacy and closed nature of online writing; for him, the challenge of having to struggle with both complexity and order in the same communicative task is one that students need regardless of the technologies they will ultimately employ, or the environments in which they will communicate.

Hesse's challenge to keep the essay as a central tenet in composition courses is helpful because it asks all of those concerned with technology to be able to argue for changes they would offer by demonstrating how much courses and the discipline itself would retain as well as how much they might change as a result of our technological leanings. One of the challenges and the opportunities of digital writing is that it doesn't have to be so stringently tied to old notions of writing. It doesn't need to be tied exclusively to writing, even. Even among those who argue that composition should still be primarily about writing, however, many would agree that it should prepare students for the challenges of different kinds of writing in vastly different spaces than just the wordprocessing programs that were the focus of so much of 1980s and 1990s thought about computers and writing. Just as one might argue that composition should continue to be primarily about writing, writing is every bit as much about technology because of the changed spaces in which writing occurs. These different spaces, like word processing programs, hypertext editors, Instant Messaging interfaces, email, chats, MOOs and MUDs, Windows Media, IMovie, Flash, and Fireworks (and the many more we could name) as Jay David Bolter has pointed out, pose widely varying demands on writers. These varied spaces demand the ability to compose just as comfortably in informal as in formal contexts, and navigate the wide ranges in between; work in synchronous and asynchronous environments; deal with both digital and print paradigms; work individually and collectively—a range of skills and abilities and understandings that demands students entering the university has some way of getting their bearings in such a mélange, and design is one of the best of the many possible navigational systems scholars and teachers have to offer them. Ideally, literacies, rhetoric, and design would operate together, equally, in composition classrooms.

African American rhetoric has rarely taken up design issues explicitly, even though ironically, the work of rhetors throughout the tradition has often been about "design" in some sense—about how individuals and groups, through language, work to transform the spaces Black people have lived, worked, played, and prayed in. Even when those rhetors were not as clear as Amiri Baraka about the centrality of technologies and their design in Black life, African American music, writing, preaching, politics, and teaching have always had as at least a subtext attempts to change the landscapes of power relations between African Americans and the rest of society. Langston Hughes' (1998) classic essay "The Negro Artist and the Racial Mountain" is a perfect example of this attempt to redesign the nation. Those some have criticized both the Black Arts Movement and the Harlem Renaissance as partial or total failures beyond their arguments about transforming Black life or the emergence of a "New Negro," Hughes' landmark essay makes this connection between rhetoric, design, and collective struggle at least implicitly through the building metaphor that controls the essay: "we build our temples for tomorrow, strong as we know how, free within ourselves" (p. 1271). The artist, in Hughes' estimation, is one who tells the stories of a people in the attempt to redesign the individual, the race, and the nation.

Beyond Hughes' implication of design as central to a conception of the artist or rhetor, there are rich traditions of African American visual culture that address the role of design in collective struggle more directly, from those who design book covers, to artists like Carrie Mae Weems, John Biggers, Romare Bearden, and Kara Walker, to Black quilters, to a long lineage of African American architects and builders going back to the first designers Booker T. Washington hired to design the buildings of the Tuskegee Institute. The tradition of Black architects and designers of physical spaces goes back even further to slave builders who often designed the buildings and spaces they later constructed. Melvin Mitchell (2001) both celebrates and calls into question this history in his book *The Crisis of the African American Architect*, again, asking why, as of yet, there are no distinctively Black approaches to architecture, especially since the living conditions of African Americans have always been a primary focus of Black struggle. Mitchell's question is the question: not just about what African Americans can do with architecture, but about what rhetoric can do in real people's lives. Design has to become a part of African American rhetorical study because it is both rhetorical and material. Sermons alone don't get houses built (although, admittedly, they might get the preacher's house built); architectural plans do, however. Songs might make people feel good, but they don't keep those people from being killed by police. The technologies those police are provided for enforcing laws can, in a system that is explicitly designed to demand and ensure that police respect the lives of those they encounter.

It is also critical that African American rhetoric engage the possibilities available in design because, for all of the conversations that have taken place about design in rhetoric, composition, and technical communication, all of those talking, and all of those talked about have been White and almost all have been male. Some of that work is certainly valuable. Cynthia and Richard Selfe's now classic "The Politics of the Interface" did valuable work in exposing the ways power and exclusion can be reified in technology design, African American students and scholars must bring to the proverbial table the ways Black struggle and history have always been about challenging and transforming interfaces, whether those interfaces of American life be classrooms, law enforcement agencies, voting booths, chat spaces, or the entire legal system including the Constitution itself. Having students begin to explore design issues—even if they are rooted in print paradigms of writing, in the essay, in web pages, can begin to denaturalize some of the conventions that they—and we—take for granted. Even seemingly simple issues like typefaces and headings, subheads, text size, and textual arrangement in courses that don't address digital concerns at all can still convey the powerful message that students are not trapped into one kind of interface. To take another Harlem Renaissance example, Jean Toomer (1988) understood this in his genre bending classic *Cane*, playing with the design(s) of literature so compellingly that one could argue it was the one work (along with Eliot's *The Waste Land* [1922]) that made literary modernism more than a rhetorical question. Critical Race Theorists like Kimberle Crenshaw, Patricia Williams, and Derrick Bell understood these connections between design and rhetorical production as well, attempting to redesign legal scholarship and jurisprudence through the redesign of the generic conventions that govern people's interactions with the legal system. Black Arts Movement poets understood these connections as well, from Amiri Baraka, who made the call explicit, to Sonia Sanchez and Haki Madhubuti, in pieces like Sanchez's tribute to Billie Holiday "for our lady" and Madhubuti's manifesto/poem combination in the *Introduction to Think Black*. These artists consciously used the page as a canvas to break conventions and create new ones, and to ultimately liberate the poem from the page, from narrow expectations of what poetry had to be, what reading had to be, what being had to be.

African American design traditions, however, whether in print, physical, or virtual spaces, have been so ignored as to seem nonexistant. The stories of Black quilters and architects as designers can help to demonstrate the crucial roles design has always played and can continue to play in collective action and day to day living—roles that open up participation to many, if not all, rather than consigning virtual futures to so-called experts and consumer-driven notions of technology, as done by William Mitchell. Those stories can also begin to point the way to answers to the challenges of people like Melvin Mitchell and "new" Black Aesthetic proponents like Arthur Jafa and Greg Tate: to find the language, the arguments to articulate a specifically African American approach to design that can bridge the many different kinds of design (from interior design to architecture, to industrial design, to web design, to curriculum design, textual design, and many others), work to undo design professions' sordid history in upholding race, and open up interpretive space for the broad, diverse range of work Black designers might pursue without imposing artificial demands on their work.

DESIGNING FREEDOM INTO THE ARTIFACT: SLAVE QUILTS AS MAPS OF THE UNDERGROUND RAILROAD

Thanks to visual artists like Faith Ringgold, Black quilters have finally begun to receive some of their due as artists and as crucial figures in African American culture. Museums now hold exhibits, and art critics offer rich analyses of individual quilts and the traditions from which they emerged. Much of this attention, however, especially in the exhibits and critical interpretation, has been limited to aesthetic analysis rather than examination of the purposes the quilts actually served and the cultural contexts from which they emerged. Jacqueline L. Tobin and Howard University scholar Raymond Dobard (2000) challenge simple stylistic renderings of the slave quilts in their book *Hidden in Plain View: A Secret Story of Quilts and the Underground Railroad*. Quilts were long known to have been an important part of the Underground Railroad, but Dobard and Tobin's work is the first to definitively describe and analyze the sign system the quilts helped to create—the ways individual stitches, images, and patterns held specific meanings for slaves, directing their actions and routes, and instructing them on how to avoid the many dangers involved in the escape from slavery. Tobin and Dobard's account suggests ways that even the most seemingly innocuous artifacts and the most routine examples of design can help counter highly organized systems of power, and help people find "freedom." The key to the possibilities Tobin and Dobard open up lies in individual designers' ability to connect their work up with present and recovered traditions, and, like the quilts, design into their tools, texts, and spaces, the sets of codes, the sign systems that can help people navigate through the hell of any physical or virtual landscape—such as systematically underfunded school systems, redlining banks, corrupt police forces, stagnating bureaucracy, a legal system that protects privilege instead of justice, coded racial assumptions and stereotypes, and severely limited access to communication technologies—into safety and freedom.

It is somewhat commonly known that slaves devised and used many covert ways of coding knowledge, both within the confines of slavery, and in their incessant attempts to escape from it. Spirituals and work songs, storytelling, dance, and blacksmithing were all connected in a network along with the quilts to help slaves endure their lives on plantations and plan and execute their escape from them.

What has received less attention, especially outside of the spirituals and work songs, are the codes themselves, and the ways those codes worked to aid slaves. Tobin and Dobard's work begins to develop these connections in their recovery and explication of an "underground railroad quilt pattern," as it was revealed by Ozella McDaniel Williams, a South Carolina quilter who held generations of knowledge about quilting traditions until her death in 1998. The quilts, made by slaves as well as quilters and visual artists after emancipation, functioned by stitching literal maps along with instructions on how to obtain provisions, decide which routes to follow, and which steps to take at various connecting points on the railroad through cultural mnemonic devices developed in Black culture through African retentions and American realities. The quilts worked so well as such a set of signs because of the ways their instructions and maps were couched in a visual language of the everyday.

The code, as Tobin and Dobard related it from Ms. Williams, contains the following instructions presented in visual patterns sewn into the quilts:

> There are five square knots on the quilt every two inches apart. They escaped on the fifth knot on the tenth pattern and went to Ontario, Canada. The *monkey wrench* turns the *wagon wheel* toward Canada on a *bear's paw* trail to the *crossroads*. Once they got to the *crossroads* they dug a *log cabin* in the ground. *Shoofly* told them to dress in cotton and satin *bow ties* and go to the cathedral church, get married and exchange double wedding rings. *Flying geese* stay on the *drunkard's path* and follow the *stars*. (pp. 22, 23, emphasis in original)

Tobin and Dobard acknowledge throughout their work the many difficulties that hinder attempts to definitively interpret the code, but they work painstakingly through research in American and African American quilting traditions, the Underground Railroad, and secret societies like the freemasons (especially Black Masonic lodge named after Prince Hall, who founded the Black lodge after applying for and being denied admission into all 150 of the lodges then operating in the nation). Through this research, Dobard and Tobin substantiate the elements of the code Ms. Williams reveals, and offer a theory of what it meant and how it operated.

Quilts made by slaves provided excellent cover for those attempting to escape because quilting was a set of skills that could be used and taught in the open, given their functional, everyday use. Many other kinds of communication, like reading, writing, and the use of any kind of drum, were strictly forbidden and even outlawed for slaves. Because slaves were under such constant surveillance for any activity perceived to be threatening, knowledge needed for resistance and escape was encoded into seemingly harmless activities. Often these were activities that played into slavemasters' assumptions that slaves were lazy and perfectly content with their condition. Elaine Richardson examines the ways slaves used song to this end, particularly in the "game" involved in shucking.

Tobin and Dobard demonstrate that more than an artifact resulting from these activities designed to trick slavemasters, the African American quilt became a "fabric griot" that employed the tactic of communicating secrets in the light of day, allowing messages to "be skillfully passed on through objects that are seen so often they become invisible" (p. 36). David Walker's *Appeal* exemplifies this relationship between design, messages, and networks of use in the service of underground communication, as well. His jeremiad was printed at a size that would allow it to be easily hidden; he sewed it into the clothes he sold in the used clothing store he owned. These examples only begin to suggest the possibilities that can result from systematic exploration of African American visual communication and design.

Aside from the specific code Ms. Williams revealed to Jacqueline Tobin, African American quilts helped slaves on the underground railroad by acting as maps of lands surrounding plantations, charting routes and distances to safe houses, including the distances between them, and advice on how to navigate those routes—all of this in ways that still inform African American visual culture. Tobin and Dobard note that the quilters "employed a visual rhythm [that] allows for no straight lines" (p. 46), not only to advise fugitives to navigate in crooked patterns to avoid slave catchers, but because of retentions from African beliefs that "evil travels in straight lines" (p. 49). The influence of those retentions resounds, from the Tate and Jafa (1998) quote about the beauty in zig-zag cornrows to superstitions embedded in childhood games. The navigation aids in the quilts used scales based on the distances between knots to suggest distances to safe houses, and colors and patterns revealed geographic features and even crops on the lands of plantations (p. 93).

The code itself, with the quilt patterns represented in the bold phrases of the previous quote, was contained in ten quilts, with nine dominated by individual patterns and the tenth, a "sampler," containing all of the other nine as a way to both tell the entire code and let slaves know which symbols would be used:

> According to Ozella, there were ten quilts used to direct the slaves to take particular actions. Each quilt featured one of the ten patterns. The ten quilts were placed one at a time on a fence. Since it was common for quilts to be aired frequently, the master and mistress would not be suspicious when seeing the quilts displayed in this fashion. This way, the slaves could nonverbally alert those who were escaping. Only one quilt would appear at any one time. Each quilt signaled a specific action for a slave to take at the particular time the quilt was on view. Ozella explained that when the Monkey Wrench quilt pattern was displayed, the slaves were to gather all the tools they might need on the journey to freedom. The second quilt placed on the fence was the Wagon Wheel pattern, which signaled the slaves to pack all the things that would go into a wagon or that would be used in transit. When the quilt with the Tumbling Boxes pattern appeared, the slaves knew it was time to escape. How long each quilt remained on the fence before being replaced is not known. Ozella suspected that a quilt would remain up until all who were planning to escape had completed the signaled task. (2000, p. 70)

Other patterns in the code specified detailed routes to take to freedom and the actions slaves should take at various places along the way. The Bear's Paw pattern informed slaves that they should follow the routes bears took through the Appalachian mountains toward Cleveland, Ohio, which was designated by the symbol of the Crossroads (p. 71).

The slave quilts, in Tobin and Dobard's retelling of how they were used as tools in the escape from slavery, offers several important ideas to an African American rhetoric of design. Aside from the near-obvious notion that design is much more than style or the aesthetic, while it might include all of those overlapping concerns. The quilts also offer some more substantive lessons for an African American rhetoric of design: that messages, even explicit instructions on how tools might be used toward liberatory ends can be designed and built into the artifacts themselves, and that the most important technologies and uses to explore are often the everyday.

ARCHITECTURE OR REVOLUTION? DESIGNING FOR THE OTHER SIDE OF FREEDOM

African American architects represent underground design traditions for different reasons than the slave quilters. Where the work of the quilters demanded secrecy, Black architects have been forced underground by the nature of their exclusion from the profession. This underground status, however, has fostered a contemporary moment in which Black architects have begun to challenge the assumptions, knowledges, and practices of mainstream architecture and the power relations it interprets and preserves.

While Melvin Mitchell laments the lack of a distinctly African American architecture to parallel other Black modernist movements like the rise of African American literature and art, he maintains throughout his book that the resources for developing such an approach to architecture do exist. Mitchell essentially argues that African American architects lost the potential they had for formulating a movement with their work when the center of architectural education moved from Tuskegee University to Howard University following the death of Booker T. Washington. Mitchell believes that the move to Howard precipitated a focus on imitating White models in the name of "an elitist White version of professionalism. That term was a euphemism for the drawing of a sharp class-based distinction between the genteel architect and the more folk and craft-based building contractor" (p. 9). A committed synthesis with these folk and craft based conceptions of design—a willingness to see those designers as important to the field—would have opened up architecture to traditions and techniques going back to slave craftsmen and an African past where design and building operated together, holistically, where builders, blacksmiths, and other craftspersons played vital roles, and would have led nascent architects to base their work more directly on the needs of Black communities, rather than merely seek access to White firms and funders.

Such a different conception of the role of the Black architect could potentially have opened those architects and their work to the visual arguments and philosophies of artists like Aaron Douglass in the Harlem Renaissance and Bettye Saar during the 1960s and 1970s, a visual-rhetorical history Michael D. Harris (2003) recounts vividly and attentively in his book *Colored Pictures: Race and Representation*. That Black architects never took up the implications of work by these artists or writers like Alain Locke and Langston Hughes is a particular failure, on the level of the charges Harold Cruse levels at Black intellectuals in the book that is the namesake for Mitchell's (2001) work, *The Crisis of the Negro Intellectual*. Cruse's thesis, attacked by some because of the vitriolic tone of his criticism, was a broadside attack on African American leadership as represented by the "intellectual." A sample of Cruse's (1984) text shows the level of invective: Black leadership "is the most politically backward of all colored bourgeois classes in the Western world. It is a class that accepts the philosophy of Whites whether radical, liberal, progressive, or conservative without alteration and calls it leadership." For Cruse, it was up to African Americans to come up with their own solutions to the problems of racism, which he defined as having material, structural, and cultural foundations. This Black activists, intellectuals, and "leaders" needed to work on what he called the triple front of politics, economics, and culture, if they are to make any progress in dealing with the effects of racism, not to mention solve its root causes.

Melvin Mitchell's solution for architecture is a triple front approach of its kind. Mitchell argues that African American architects can better serve Black people and Black communities by pursuing a New (Black) Urbanism, which for him consists of:

1. focusing on African American communities, and particularly homes instead of relying so heavily on public buildings to sustain their practices,
2. understanding the role of the Black architect as one of a combined "designer/builder," in which architects become more self-sufficient by managing all aspects of the process from planning to construction, and
3. looking to Black cultural traditions for the ideas, materials, and even designs upon which to base their work.

Mitchell's solution addresses political, material, and cultural realities and proposes a transformation of what architecture is and can be that clears intellectual space and opens new audiences for Black design work. It also reduces dependence on the networks of power that have resulted in such rigorous exclusion of African Americans, women, and other people of color. To cast it in different terms, Mitchell proposes a transformation of African American architecture by preserving the underground, and making that cultural underground the space of engagement, a space where African American futures can be dreamed, imagined, planned, and designed.

There are architects and other design professionals who take up Mitchell's challenge, attempting to move toward a specifically African American aesthetic.

Miles Davis and Josephine Baker are often cited sources for such an aesthetic, both of them credited with having changed the ways lines, angles, colors, and absence can figure into design. Craig L. Wilkins (1998), in his article "A Style That Nobody Can Deal With: Notes from the Doo Bop Hip Hop Inn," issues his own call:

> The development of architecture for this new temporal-spatial community as a form of oblique challenge to structural domination would surely be what LeCorbusier, that most voracious devourer of the "other" in the name of hegemonic spatial appropriations, argued against when he boldly declared in *Towards a New Architecture*, "Architecture or revolution. Revolution can be avoided." If the current, demoralizing state of our urban environments are any indication of the position of "architecture" in Corbu's dichotomy, I'm opting for the revolution. (p. 26)

Wilkins sees Hip Hop, particularly as Tricia Rose understands it in her (1991) *Black Noise: Rap Music and Black Culture in Contemporary America*, as an important part of the response:

> Having taken up the challenge posed by Corbu, I reject the historical proposition that architecture "will always look like architecture." The architectural discipline is guilty of complicity in the systematic marginalization of people. Showing us that marginalization is no longer inevitable, if it ever was, Hip Hop has taken dated technology and devalued material from the trashheap and made a global culture out of these discarded parts of a postindustrial urban environment. Inspired by Hip Hop's aesthetic of layering, rupture, and using existing, and even discarded materials in new and creative ways, under-capitalized architects of the people can nevertheless be well on our way to developing revolutionary, relevant style through what Carl Boggs calls a "new code of space." (p. 23)

Wilkins understand music, and Hip Hop culture specifically, as refiguring lived spaces through the ways they refigured imaginative spaces. Rhetorical acts involving the taking, the jacking of access *and* a refusal to be satisfied with any kind of limited access come together in acts of design with the potential to intervene in long-standing systems of exclusions and possibly even refigure social, economic, and political relations for African Americans and others who have been subject to those exclusions.

CONCLUSION: TOWARD AN AFRICAN AMERICAN RHETORIC OF DESIGN

Just as Hip Hop could not have created this potential without the African American cultural underground, neither can design. Even with rich underground spaces and traditions to draw on, Black designers have found it difficult to develop and articulate a specifically African American design aesthetic. Participants in the 1994 Organization of Black Designers conference reflected this difficulty in an attempted definition of a Black aesthetic: "the Black aesthetic must be inclusive, not exclusive. It must have a natural appeal to humanity and be a synthesis of many different forms. It must be dynamic and participatory. And most of all, it must have soul. This is what sets it apart from all other levels of design" (p. 10). Clearly, there is much of value in this initial attempt at a definition, specifically, its openness to a multitude of forms and its insistence on participatory approaches, as opposed to the closed, system centered, expert-driven approaches that undergird William Mitchell's and LeCorbusier's conceptions of architecture. One is left to wonder, however, where the "soul" is in the description. Aside from the word soul, what is there in the description that carries the mark of African American experience? Part of the difficulty lies in the process of defining concepts like that, concepts that are so ingrained in African American experience that once one attempts to pin them down, one is hopelessly lost, seemingly condemned to vagueness.

Definition is not impossible, however, of "soul," or of an African American rhetoric of design, as the examples of Black architects and quilters demonstrate.

A rhetoric of design that takes ideas from soul, Hip Hop, the blues, jazz, the worksongs, and spirituals as well as the visual cultures that accompanied those forms—a framework for the visual principles and arguments that can be engaged and employed in the pursuit of justice for Black people while preserving and celebrating African American culture—is also possible to define, and based on the examples of Black quilters and architects, would consist of the following:

- a move beyond the word in African American rhetoric, to see design practices as an important element of larger struggle, of offering ways to resist the stubbornness of racism and racialized exclusions;
- a commitment to charting and maintaining the continuity of specific visual patterns and practices, ideas, and materials, from the admonition to travel in crooked lines because evil travels in straight lines, all the way to the zig zag cornrows, asymmetry, and contrariness of the present;
- a participatory approach to design that remains committed to developing and maintaining underground spaces for design work while also interrupting the larger patterns of exclusion of design professions and discourses—an approach that engages lay people in the act of redesigning the textual, physical, and virtual spaces we occupy rather than submitting to experts who are complicit in those exclusions;
- an approach to design that encodes freedom directly into spaces and artifacts, in a sense, an approach that embeds the technical communication, the instructions, the documentation, into the artifacts and spaces themselves;
- a willingness to use every means available in design, even dated and discarded technologies, and spaces and artifacts of the everyday; and

- an approach that assumes intellectual and physical freedom, no matter what hindrances still seem to prevent that freedom.

The lessons of Black quilters and architects, encapsulated in these principles for an African American rhetoric of design, show that not only is such a rhetoric of design possible, but that design can be an important site of the continued pursuit of justice and inclusion of African Americans. Just like preachers, teachers, politicians, poets, and others who work words toward the ends of equality and a meaningful access to the spaces and technologies that make up American society, designers can have crucial roles to play in leading people through the hell of racism and race-based exclusions into freedom, no matter how they might define it.

7

A Digital Jeremiad in Search of Higher Ground: Transforming Technologies, Transforming a Nation

In some ways this book is a kind of digital jeremiad, or at least a response to the call of Black jeremiahs throughout our history. Martin Luther King, Jr. warned us nearly 40 years ago. Many of us slept as the signs changed telling us that the world we knew had changed from an industrial one to a digital one. Many of us who were awake might as well have been asleep subject to only the most BASIC of understandings about what technologies could do, could be, the purposes to which we would put them, and the relationships we would have with them. The broader challenge is exactly as he identified it in 1968—the challenge of transforming our technologies as we gain access to them, and that we use those technologies toward the larger project of transforming the nation, of justice and equal participation for Black people and all people. The truth that has been the bedrock of the African American jeremiad and the reason for its central role throughout African American rhetorical traditions remains: either we accept the hard work and sacrifice of this challenge or destroy ourselves in attempting to run from it.

Access will no longer do as the excuse for limiting ourselves and our students to BASIC writing, BASIC literacy instruction, BASIC technology instruction, BASIC diversity efforts. Transformative access as the ONE in African American Rhetoric and as the ONE in rhetoric and composition and as the ONE in technical communication and as the ONE in computers and writing and as the ONE in science and technology studies and as the ONE in Africana studies—the ONE linking dialogue within and across disciplinary lines offers rich possibilities for moving ourselves, our students, and even our technologies and society out of our current

Digital Divide, out of our centuries-old racial ravine, and onto higher ground by addressing concepts each of these areas already embraces in one way or another: the related axes of critique, use, and design, employed in an ethos rooted in the jeremiadic vision for a society worthy of the ideals that brought us all here. King's version of the jeremiad and his vision connected technologies, race, and the future of the nation in his favorite metaphor, his insistence that we are all bound together, like it or not, in an inescapable network of mutuality. King argued passionately from the Montgomery Bus Boycotts that forced him out of post-graduate school naivete and onto the national stage and into a history of struggle to the days before his death working on behalf of striking sanitation workers that our nation's health, that its survival depends on our ability to transform our technologies and ourselves from tools, processes, and systems that uphold domination and exclusion to instruments of justice, participation, peace, and love.

My favorite metaphor for the search for that transformative vision that unites King's speeches, Malcolm X's television interviews, anonymous slave quilters, recreational chat users, critical race theorists, and millions of unnamed leaders, laypeople, teachers, poets, politicians, preachers, parents, students, designers, journalists, entertainers, and others is the search for higher ground. Despite our common sense of the importance of Parliament/ Funkadelic's conception of Funk—especially as it is expressed in the ideas of the Mothership and "The One," I do not share Robin Kelley's belief that PFunk stands as a musical embodiment of the radical or revolutionary, particularly when I search for models of rhetorical and technological innovation for all our students, and particularly African American students and students of color. There are clearly "freedom dreams" and radical potential in PFunk's concepts. There is also the same potential in the forces that gave birth to hip hop as a movement, although many artists seem to have abandoned that potential with their uncritical embrace of the industry and the networks of power that determine radio airplay and video rotation as they invoke a politics of authenticity by claiming they are "keepin' it real" and "representin" the truth of Black experience.

But Funk's sometimes nice, sometimes nasty elder sibling Soul probably provides far better examples of what the ONE can be in African American rhetorical study. When it comes to the combination of radical imagination, rhetorical and technological mastery and innovation, and the commitment of all those gifts in the service of one's freedom dreams, the example from Black music (because, after all, we are a "blues people," as Amiri Baraka reminds us) who has always kept it real, represented truth, and produced on his radical potential is Stevie Wonder. His contribution on all of these different levels serves as a model from popular culture of what African American rhetoric—and Rhetoric and Composition—can be. Beyond the exemplar that Stevie was and still is, however, the Soul era provides the ideological range that bridges Civil Rights and Black Power, the stylistic diversity that connects unphased cool and the intense heat of passionate insistence on justice, and the spectrum from participating by playing by the rules to throwing those rules away and still demanding to participate to outright resistance of spaces and systems that have excluded us. The combination of gospel vision and blues sensibility—of the demand for immediate results and a focus on the longgame (shout-out to Arthur Flowers for the term), of the demand for equal access and insistence on transformation—that underlies all of African American struggle in some way is exactly the ethos that should guide our encounters with technologies and the exclusions that have marked them.

What is Soul? What kind of intellectual work does "Soul" do as a concept? How can it inform technology use, critique, and design? Soul, as Stevie Wonder understood it in that call to search for higher ground, Soul as Blackfolk understood it and still understand it, Soul as it was and is embodied from Sam Cooke to Curtis Mayfield to Etta James to Marvin Gaye to Aretha Franklin to India Arie to D'Angelo to Jill Scott, is something far, far more than the mere essence or spirit of something. That spirit, that essence, that ethos, is a part of the answer, as Manning Marable tells us "from the interior of our own being, it is an ethical foundation for the choices we make in our lives and the sense of responsibility we feel in how we relate to other people. Soul also implies the possibility of transcendence." Marable brings together survival and victory, struggle and strength, going on to say that "soul also implies memory, agency, and hope in the face of despair. The Souls of Blackfolk have encountered terrible exploitation and inhumane conditions, a degraded and desperate physical reality that would force most reasonable people to conclude that God, or whatever one might call universal truth, had abandoned and forsaken them. In the construction of Black culture, to have soul is to be truly at home with oneself and with the people. Soul helps us to navigate the hostile currents of an unequal and unfair world, a world stratified by color and class, where all too frequently there seems to be no justice." But even more than the seemingly absurd faith that African Americans found in themselves and in a nation that has continually taunted them with the refrain "we hold these truths to be self-evident," soul is participation and resistance, substance and style, access and transformation.

But Soul is even more than that.

Pride. Roller skating parties backyard barbecues a willingness to struggle together and share whatever we had or didn't have. The house party. Angelic voices singin the devil's music with all the passion power and commitment of CL Franklin's preaching. The slow jam. The ONE. Horn sections so tight they snap like snares. JBs extralinguistic grunts and babblin. Barry's beggin. Etta's pain. Donny's melancholy, his take you to aplacewheretheresnospaceortime interpretations. Marvin. Stevie. Streetcorner harmonizers imitating the delfonics or the dells. Ray, Goodman, and Brown, Take 6, and BoyzIIMen keepin it alive. Jill Scott callin us out, then tellin us keep fightin warrior I know you're there. Blackademics—Geneva Smitherman, Keith Gilyard, Jacqueline Jones Royster, Bill Cook, Rosentene Purnell, Jim Hill, Vivian Davis, in the 70s and 80s doin the damn thang and openin doors. Forcing an entire discipline to change. Afros. Amiri the prophet tellin us look we can make technologies look like US, do what WE want them to do. braids. That unmistakable walk with just enough lean, just enough bop. Gordon Parks catchin just the right moment at just the right angle at just the right time to tell the story of a people in one shot. The perfect cool projected in resistance to centuries of struggle—and the dogged determination to keep on pushin. Shirley Chisholm, unbought and unbossed. An idealism purchased in blood. The hope that this society can finally get it right and the insistence that it do it NOW.

Soul is Coland Leavens at Cleveland's Charles Eliot Middle School teaching students to master Standardized English and rhetorical forms like no one else can *and* keeping the languages, the Ebonics, the Spanglish, the indigenous languages of home alive. Soul is the ability to recognize complexity and still pursue unity. Soul is understanding that that one goal is something far far more than Diversity or Multiculturalism, but racial justice. Soul is knowing that the occasional black face, or person of color does not mean justice for Latino, Native American or African American people. Soul is knowing that not only is racial justice *the* question, but that its answer does not lie in some tired debate over personal responsibility versus government intervention, but about a larger job of transformation. Soul knows that the justice that is still denied so many is about organized systems of power, White privilege so programmed into our daily interactions and political interfaces that people swear it no longer exists, about histories of lies and misrepresentation, and about economic injustice that feeds the fire of old racist exclusions.

Soul is what you see happen at a gathering of Blackfolk when the DJ plays Al Green. Soul is the challenge to think critically that was underneath Marvin Gaye's classic "What's Goin On?" and "Mercy, Mercy Me." Soul lets us love each other even as we insist that we do right, like Aretha did in RESPECT and Jill Scott did when she said "keep fighting warrior I know you're there." Soul is technical mastery and the freedom to completely discard the rules that bind us. Soul is Dr. King's vision of the Beloved Community, the integrated society we create where there is justice for everyone, respect for everyone, concern for everyone, *and* it's Malcolm's vision of healthy Black communities understanding the distinct beauties of the culture we have forged in this fire of struggle. Soul is when liberal and progressive and moderate and radical (and even some conservative) Black people woke up and reached across ideological divides because their minds were stayed on freedom. The SOUL era represents the best of who Blackfolk were and the best of who we all can be because African Americans marched, fought, protested, worked and prayed for more just economic, educational, political, and social inclusion for themselves *and* demanded those same doors be opened for everyone, including many White women who walked through them during the 70s and 80s and became feminists who forgot about race and soccer moms scared of Negroes.

Soul is the continual, committed search for higher ground, whatever the limits of public or scholarly discourse might seem to be.

What would happen if the Soul aesthetic embodied in Stevie's search for higher ground were the aesthetic that guided technology use, critique, and design? What would happen if writing instruction found room to honor the flair of Dr. J's classic come-from-behind-the-backboard scoop and layup? What would happen if the Freedom Schools organized as a part of the Mississippi Freedom Summer became composition's and technical communication's curricular model? What would computers look like if they carried the spirit of Aretha Franklin's chord progressions when she sang "I Never Loved a Man." What kind of access to technologies would African Americans and people of color have if our digital literacies were organized around those Malcolm and Martin exhibited in their debate about and activism for access to the franchise?

The notion of a transformative access can be the downbeat, that One, that can show students, teachers, and activists we do not always have to choose between rhetoric and technology, between rhetoric and the real, between analysis and production, between participation and change. The lessons toward achieving that One provided by the brief look at the texts examined here, are several. The first is that any meaningful access must be about more than the material, or one's presence in a particular space. Regardless of the number of African American lawyers practicing before the bar, the same presumptions about Black people that A. Leon Higginbotham describes ensure continued disparities in how African Americans are treated at every level of the legal system.

Access requires an individual or group of people having the material of any particular technology, along with the knowledge and experience and genuine inclusion in the networks in which decisions are made about their design and implementation that enable them to use—or refuse—them in ways that make sense in their lives. Combining those four levels of access (material, functional, experiential, critical) in some way that can represent transformation is similarly a multi-faceted task. People must think and act simultaneously along the axes of critique, use, and design. The Jeremiah cannot work alone to point out the failings of a society; people must also be prepared to imagine, design, and build new systems, new documentation, new tools, new networks that assume and naturalize the epistemologies of those who (in this case, African Americans) have been left out. Their histories and assumptions must be naturalized, centered, in those new spaces. And finally, along with both critique and design, those who press for change must be able to count on users to participate in those new, technologized spaces, as problematic as they might seem. The protests of Black students to open up access to higher education in the 1960s and 1970s would mean little today if there were not faculty and staff who confronted the ugliness and the ignorance of those systems in the service of Black students—even when their participation in those systems does not necessarily represent the most radical critique or redesign. There is no need to choose between Malcolm and Martin; the vernacular traditions of Black quilters or the aesthetics of hip hop; recreational uses of computers or serious academic or technological work; undergraduate student protesters or the wizened souls who seem quiet but helped those undergraduates get admitted in the first place. Without any one side of the equation, the other equals zero as well.

Taken in the context of the rhetoric and composition classroom, there must be public and private phases of struggle; exemplars and lay users, lay texts, and entire traditions, argumentative and expressivistic writing, clear standards thoughtfully applied and ungraded spaces where students can experiment and figure out who they be without sanction. On a technological level, both the newest advances and discarded materials are important in all of our attempts to find our collective way home.

If we truly paid attention, neither technical communication, rhetoric and composition, nor African American rhetoric could be the same in the face of the challenges posed by the Digital Divide's ten year history. Funk and Soul and African American traditions from slave quilters to Critical Race Theorists to Martin and Malcolm to lay users of BlackPlanet offer one set of answers that I've documented here. Robert Johnson (1998) offers another. Drawing on a range of technical communication scholars who have attempted to recast Aristotelian ethics in a social justice light, Johnson argues that no matter how overwhelmed teachers might feel by all we are asked to do, "acting through" our knowledge is central to our roles as rhetor and educator. The action that Aristotle's ethics demands, in Johnson's view, is central to the "development of a technical rhetorician: a technical communicator who is trained in the theory and practice of the arts of discourse *and* who practices these arts as a responsible member of a greater social order" (pp. 154–158). Johnson's attempt to move technical communication away from system-centered theory and toward an approach focused on users is one that aims to interrupt the assumptions and practices that have produced such rigorous patterns of domination and exclusion in our technologies.

Johnson's attempt at changing technical communication is a bit understated, however, because it does not allow him to consider those patterns of exclusion directly or imagine what non-users and new users who have been systematically excluded might to the tools, the techniques, the theories that make up our technologies. What would computers look like, what would they do, if technology companies and researchers really studied African American computer use, and designed computers attentive to their findings? If rather than look for one software package, one design aesthetic at a time for cases, monitors, interfaces—if rather than look for one standard of everything, those designs were culturally relevant for African Americans and other groups of people. What would computers be and do if individual users and those in the technology sector took Amiri Baraka's arguments seriously? If Martin King's demand that the only test for our technologies is the extent to which they foster peace and improve people's living conditions and relationships?

So ... if Stevie Wonder or maybe even George Clinton were rhetoric and composition scholars and/or taught technical communication, we just might have a chance. Because that's obviously not the case, we have to talk to each other across the fractured lines within our discipline and across disciplines to refigure our teaching and research to make questions of technology access central to our work. What follows, without attempting to prescriptively offer a specific agenda, are some thoughts about what teaching and research projects to that end might look like if technology access were to be seen as the major ethical issue for scholars at the intersection of technologies and writing.

TECHNOLOGIES AND WRITING INSTRUCTION: MAKING ACCESS REAL

Anyone who works in a school knows that when a news report proclaims that "over 90% of all schools in the nation have access to the Internet," that number gives absolutely no indication of whether or not students experience computers or Internet technologies in meaningful ways. We know that kind of number is no measure of the percentage of schools or students who have embraced technology. We know it doesn't get even close to the percentage of schools, universities, or teachers who have implemented carefully thought-out technology plans or technology literacy curricula.

But that 90% (or even more in some reports) is still given as a magic number to suggest that students get an equal opportunity to work with and learn about technology. This fraud is also committed when policymakers decide that because a public library has 6 computers, that the 6,000 people in that community it serves have access to those computers. Or that because a community organization got a grant to try to minimize the digital divide, the computers it is able to purchase actually close that canyon.

These schools, libraries, and community organizations don't commit this fraud. Teachers, librarians, community advocates often want desperately to serve their constituents, and understand the importance of computers and the Internet as tools that can make a genuine difference in their lives. The fraud isn't the fault of all of the policymakers either—many of them have worked diligently in the last several years to put the problem of the digital divide on the table for public discussion. Nor is that fraud primarily the problem of teachers who must try to learn tools, techniques, and concepts for themselves even as they are expected to be technology experts for their students.

What is fraudulent is the definition of access that begins all of these conversations and directs our actions on behalf of those who do not have it. Just as the right to vote alone does not ensure that people have access to the local, state, or federal governments that are supposed to serve them, just as school desegregation alone did not magically provide equal access to education, a few computers and Internet connections alone will not mean that people in rural communities, poor people and other people of color will wake up some fine day and marvel that they now have equal access to technology and information.

Access to technology means so much more than the presence of a particular tool, and definitions of access that do not acknowledge how complex a problem it is are, in fact fraudulent, and will not serve to do anything meaningful for people who have consistently been denied anything close to real participation in our society. The issue for Composition teachers is how can we understand access better, and how can we work toward meaningful access for ourselves and our students. The following sections begin to answer these questions and offer specific strategies teachers can use.

BEYOND THE BOX AND WIRES: COMPLICATING ACCESS

I attempted in the first chapter to provide an outline of what might constitute a more meaningful access to digital technologies. Of course one has to own, or be near places that allows her or him to use computers, software, Internet connections, and other tools in order to have access to those technologies. I called this *material access*. But for one part of access to have any effect on people's lives or on their participation in the society, people must also have the knowledge and skills necessary to use these tools effectively (*functional access*). They must be connected enough with those technologies that they use them (*experiential access*). They must also understand the benefits and problems of those technologies well enough to be able to critique them when necessary and use them when necessary (*critical access*). We must know how to be intelligent users, producers, and even transformers of technologies if access is to mean anything to our individual lives, the lives of our students, or those of the communities we live, work, and play in.

That template for what meaningful access requires doesn't go far enough, however. What is meaningful access to technology if it's not just about its availability or proximity to us? It's not just a neat list of material access, functional access, experiential access, and critical access. Access to any particular technology occurs only when individuals or members of a group are able to use that technology to be able to tell their own stories in their own terms and able to meet the real material, social, cultural, and political needs in their lives and in their communities. Access to a technology means that members of a particular group know how to use it for both participation and resistance. Real access goes far deeper than the passive consumerism that drives almost all computer advertising and much technology policy—it is about the ability to use computers and the Internet as a means of production too.

For writing teachers that means helping students to see technologies as equipment and processes and systems of knowledge that they want to learn how to use and how to critique and how to design and create. It means teachers and students need to be involved in the messy arguments around technologies as much as in the tools themselves. For instructors who feel new to technology themselves, however, that can be a daunting charge. Rather than be frustrated by that charge, however, writing teachers can feel empowered by it, because it can connect us again to what our students go through in our classes.

Students coming into the composition classroom, no matter what their linguistic heritage, enter being asked to look at reading, writing, and thinking in very different ways than they had been asked to before. They have to pick up new languages for all of these things, and new relationships with the language varieties and discourse patterns we try to help them become more proficient in. As teachers rooted in the humanities, we can use our late entry into technology issues to share that process of being forced to acquire new literacies and skills. Where many in Composition used to (and still do) look at students using African American English or Spanglish or other varieties as weaker writers than students more comfortable with standardized varieties, we can see them as being in the same situation we're in. In other words, no one would look at writing teachers as less gifted or having less to contribute simply because we have not had the chance to become fully immersed in technology conversations. Committing fully to integrating technologies in the classroom can put us in touch with the natural awkwardness and adjustments that come with picking up new skills and acquiring new discourses, and help us to move beyond some of the debilitating assumptions we make about all student writing, and especially that of students from different linguistic traditions.

The question then becomes how to take that perspective into the reality of integrating writing technologies in the classroom. The following suggestions can help those who don't feel comfortable leaping headlong into the fray through the nuts and bolts, bytes and wires:

- *Start slowly.* You don't have to make the jump to a fully technologized classroom immediately. Different levels of technology use work for different teachers, different courses. Work at a pace you feel comfortable with. Your first step might simply be to post your syllabi on the Web; that might be enough for a given term.
- *Only use technologies to meet your curricular goals.* Never feel pressured to use a particular tool just because it seems like everyone is using it. When the tools available drive your pedagogy, disaster can follow. Think about your most important goals, then about how technologies can support them. If, for example, improving student discussions is most important to you, it might be worthwhile to introduce electronic chat from time to time. If you are concerned with the amount of writing students get to do, and journals don't sustain their motivation, you might want to explore web logs (blogs) or message boards. If you value class presentations, introduce PowerPoint or other presentation software. No matter what you use or don't use, technologies are only worthwhile to the extent that they help you create the class environment you want for your students.
- *Let your students teach you.* Better yet, make them teach you. This is the best way you can become more familiar with different tools and the ways students actually use them. Use your class assignments to have students explain to each other (and, therefore to you) how to use a particular tool. Use class de-

bate or exploring an issue assignments to talk about messy technology choices or policies.

- *Don't be scared of recreational uses of technologies.* When students want to chat, or use Instant Messenger software, or surf the net, we instinctively worry that they will just play around, and won't learn from those activities. Sometimes this is the best way for them—and you—to learn. Everything you design into your course, from writing assignments to the activities you use to set them up, does not have to be a high-stakes enterprise. Take some of the pressure off of your students and yourself when there is a chance to.
- *Don't just produce customers.* Teach your students to use technologies, but also to question them, recreate them, adapt them to fit their needs, and ultimately even transform them. Help them to look at the assumptions that lie underneath technology designs, functions, content, and marketing. Bring the critical eyes we've been trained so well to use to technologies too. This might be the best thing we can do for our students.

My own journey through these issues is still and has often been marked by the same uncertainties about what to do, how to do it, and to what ends that any digital newbie would bring to his own work and to the classroom. Despite the problems this uncertainty would seem to cause, I believe that this troubled space is exactly where the promise lies for our field—intellectually, practically, politically, and collegially. Faculty in both technical communication and rhetoric and composition programs teach from the same exigence, regardless of what seem to be major differences in pedagogical goals. Both areas place faculty in two constant and almost ridiculous binds: having to choose between "The Word" and "Technology," between being and remaining true to a developing tradition of critique and providing students with the means with which they can gain access to the university and the workplace.

These seeming contradictions, between realities of access and transformative ideals, are exactly where the challenges and rewards lie for me in writing instruction, as few other academic areas are forced to deal with them. Teaching students to communicate in widely varying physical and virtual spaces, then, must be about both teaching them what they will need to succeed in those spaces, and the traditions of critique across several cultural and political contexts that have led many to attempt to transform them. The commitment to stand in, and wrestle with, these contradictions leaves me attempting to synthesize design, literacy, and the rhetorical in ways that are not always comfortable.

I've attempted a few courses that deal with technologies in the rich ways I describe, from a first year writing course on the Soul era I taught in 2004 that asked students what a Soul aesthetic might contribute to technology design and what soul politics and activism would do to technology policy conversations, to a technical communication course I designed that was organized entirely around the Digital Divide as a rhetorical problem, examining technology policy, the politics and discursive conventions of reports, interface design, documentation, and technology use and usability. Whether students will respond well to such a course, I don't know yet, but I do plan to find out. While still in graduate school, however, I brought all of these issues together in a class I taught last summer for First Year students entering science and engineering programs at Penn State. In a six-week course, I was asked to address everything from writing skills to academic survival skills to cultural and social issues Latino/a and African American students would face while at the university. Of course, those who asked me to teach the course had little idea how such a course should look, or even, what "skills" these students would need. I called the course I designed "Transformative Access: Race, Writing, and Technology," later borrowing part of my dissertation title from it. We read across all kinds of disciplines, from architecture to NSF statistics on minority graduate degree earners in engineering and the sciences, to Alondra Nelson's (2001) edited collection, *Technicolor*, that examined everyday uses and transformations of individual technologies, regardless of larger problems of unequal access.

I framed the course with a first week in which we read Victor Villanueva's (1993) *Bootstraps: From An American Academic of Color* and Keith Gilyard's (1991) *Voices of the Self: A Study of Language Competence*, pushing the students to understand communicative excellence as knowing the conventions of every space in which they want to communicate well enough to employ and change them as they see fit. Villanueva and Gilyard helped me to emphasize the need for the students always to feel at home in language, with technologies, in their majors, and at Penn State. I wanted them never to see success in any of those places as a matter of having to choose between individual identity, cultural identity, and academic or professional identity.

Keeping in mind the fact that my students were in classes all day, 5 days a week with less than a weekend to celebrate their high school graduations I wanted to keep the course fun even as I kept it rigorous. To that end, and keeping with my continued focus on rhetorical and technological access and transformation, I made the capstone assignment of the course a technology redesign project: a lowrider computer. The legacy of Chicano/a, Latino/a, and later African American youth in completely changing the form and use of cars and car culture became a metaphor for what I wanted them to do with computers, and by extension, their experiences in their majors and the university. In addition to redesigning a computer from the processor out to the case, I required that students make the language of their essays describing the redesign match the redesign itself—their new computer's style, the creator's individual and cultural identities.

As a result, for 6 weeks, at least, students were able to play with languages and technologies, participate in them, change them, and come to voice. They were challenged. They were taught what their Composition courses would expect of them. And they were able to do all of this in a space that embraced who they were—even as they still worked to figure it all out. One student's redesign took the hyperbolic tendencies found in African American and Latino/a popular cultures and used that theme to reverse the trend in many technologies toward miniaturization, designing a computer that was a large and wonderful status symbol taking up an entire room of a house, but all made up of "vintage" spare parts. Green design and style and technology and African American, Latino/a people in the conversation—imagine that.

Whereas the technology transformation project is one of my favorite assignments to emerge from this struggle to address technology issues in meaningful ways in my teaching, there are others that I use regularly as well: one I call the "intellectual mixtape," in which I ask students to compile a soundtrack to the ideas we've covered during a particular course or course cycle. This assignment serves not only as a review exercise, but forces students to engage the ideas again and connect them to music, artists, and traditions they enjoy. This assignment also allows me (sometimes) to directly engage issues of intellectual property and ethics with digital technologies and writing processes. There is potential I have not yet tapped in an assignment like this, as it could allow one to move from basic issues like what copyright law allows as fair use to how groups have organized in resistance to the domination of major media companies in the forming and enforcement of those laws, eroding almost all sense of public good or fair use. Such examinations could then, in a technologically focused course, link various copyleft and open source movements (and their own ironic racialized exclusions) to the patterns of textual borrowing and building that marks forms like the African American sermon to such an extent that preachers often jokingly relate their citation system. Michael Eric Dyson (2000) relates the joke in his book on Martin Luther King, Jr. *I May Not Get There With You*, the first time one uses an anecdote or particular phrase, she or he cites the source directly; the second time, the preacher might say "someone once told me," indirectly citing the source and moving to the story or phrase; but by the third time using that sermon or telling that tale, the preacher incorporates it fully into his or her text, with no attribution. My goal here is not to chronicle everything I do to incorporate technological issues into my teaching, nor is it to use my own teaching as somehow exemplary. I definitely have my share of struggles. I only use those two assignments to begin to show the ways we can use specific technologies and inquiry into larger technological issues to move beyond merely functional uses of those technologies to work across the whole spectrum of critique, use, and design.

A TECHNOLOGICAL AGENDA FOR AFRICAN AMERICAN RHETORIC

A serious consideration of the relationships between communication technologies and rhetorical production throughout African American history can dramatically reshape African American rhetorical study. Such consideration should be central to the field as we reimagine it in a new century, where it can open up new spaces for inquiry, even while giving us new ways to help our students and the public appreciate how even African American language and cultural productions have been used to respond to and intervene in technological systems. It can also help to ground broader rhetorical and technological theory, both of which have simply dismissed African Americans as untheoretical. The most important task is to bring African American traditions, experiences, bodies to the technologies we use—to see technologies in our own image and never accept the default constructions of others as the only point of access, be that access material, functional, experiential, critical, or even transformative. The opportunity that such an approach opens up is, as I see it, connected by four vital tasks: a need to reexamine those works and communicators who have been a part of African American rhetorical study (the wide range of texts from poems, plays, sermons, essays, pamphlets, newspapers, journals, and more) that asks how a strong reading of those texts as technological, as well as linguistic and argumentative performances, can help us appreciate the language and persuasion in them even better; look for ways to open up what we think of as the African American rhetorical tradition, to pay just as much attention to image, body, and design as we do language in attempts at persuasion; to pay careful attention to the relationships that exist between race and emerging technologies so that we can teach students how to operate with those changed, and changing same, spaces; and finally, to integrate the work we do as thoroughly as possible with the technologies that are available to us now, to make the Talking Book really talk. Some brief examples of what these tasks might look like follows, with discussion, taking the last one first.

Digitize the Tradition as Thoroughly as Possible

African American rhetoric as a field of study should make the Internet and multimedia publishing spaces for reclaiming the work of as many speeches, sermons, flyers, posters, radio broadcasts, websites, television shows, films, and other media as possible in hypertexts and other instructional media that are collaborative, connected to each other, and in DuBois's standard for Black Theater that was reduced to a sample and turned into a clothing industry, FUBU—for us, by us. Using cyberspace and multimedia products like CDS for serious archival work can prevent those who teach African American rhetoric from having to worry as much about what publishers are willing or unwilling to publish, or include in anthologies. It can also, to some degree, help to minimize the problems of copyright that can emerge in our attempts to make materials available to students and colleagues. My point here is that if as much material as possible that is currently in the public domain and/or can be obtained through permissions is archived in repositories that make free and open access to them their policy, the struggles that individual faculty and students face in trying to obtain those permissions from the exclusive archives Alkalimat mentions, or from dead publishing companies who still hold the rights to so many documents, can be mitigated.

Examine the ways race operates online and in other communication spaces: not just the ways African American people or race or difference is constructed as Nakamura begins to do, but the ways African American rhetorics are appropriated and/or silenced. Just as important a strain in this project, however, is understanding the ways people in the divides we talk about participate in, respond to, revise, and even resist the discourses they enter. Do the ways African American people communicate on the Internet suggest major challenges on the horizon for how we define and study African American language, communication, and rhetoric? Do they provide affirmation? Some complex combination of the two? Exactly how are the ways they communicate connected to the traditions that Smitherman, Gilyard, Cook, and other contributors to these volumes chart? In what kinds of online spaces are our connections to these traditions strongest? Where are the flatted fifths—the blue notes—online, those single notes, those individual utterances that talk back to, signify on, and ultimately transform entire discourses? How do we want to teach our students to maintain or engage those connections when they communicate online?

Beyond Black language traditions, what do those uses teach us about technologies and technology use? What happens when we force technology researchers and makers to take African American technology use seriously, for its own sake, and in comparative/contrastive analysis with other groups? We need empirical studies, from the case study to rich ethnographies to longitudinal studies of what happens with writing in technologized environments—studies that also seek to understand the real effects of systematically differentiated access to those environments.

Reexamine the Tradition With Access to Communication Technologies as a Major Trope

This reexamination can allow students and scholars in the field to reclaim far more of the tradition and engage it more fully: the genius of Oscar Micheaux can be studied in light of all of the technological mastery that was involved in his development as a filmmaker along with his ability to challenge representations of African Americans on film. That technological mastery and rhetorical use of it can be connected with Spike Lee's use of digital video in *Bamboozled* as a way of circumventing the film industry's continued silencing of African American filmmakers. We can build on Deborah Atwater's examination of Stevie Wonder's political commitment and look more carefully at what was at stake when he made the transition from "little Stevie" on just his harmonica to a maturing artist who was also a technological forerunner with his choice (or Miles Davis's, or Herbie Hancock's) to use electronic instruments in an era when many in his audience were still acoustical purists—or what it meant for all of Parliament/Funkadelic's music to be organized around the metaphor of the "mothership" and a rhetoric of "the one" in an era contemporaneous with the latter years of the Black Power Movement during the 1970s.

The ways certain technologies seem so obvious to us that we don't think of them as technologies, as in the case of the soapbox, or the megaphones that were attached to cars to carry messages throughout African American communities, might be one reason this project has not been taken up more fully already. Even Black oral traditions are technological, in the sense that they users manipulating systems of knowledge through particular processes and techniques in order to communicate, from African griots to slave preachers to politicians to contemporary spoken word artists, (to steal a riff from Nikki Giovanni when she says "I'm so bad even my errors are correct!") even our most non-technological moments is technological!

Open up African American Rhetorical Study to Reclaim the Image, the Body, and Design, While Continuing to Appreciate the Importance of "The Word." Opening up the scope of African American rhetorical study in this way can not only help us consider the inordinate number of things in the preached moment that contribute to the sermon's importance to African American rhetoric: the preacher's physical presence and booming voice, as in the case of Vernon Johns, or the design of the order of worship in a church service to create drama and heighten expectations leading up to the sermon (for those interested in performative rhetorics), but can also provide us with ways to examine similar issues in entirely new spaces, such as African American community Web sites or African American community e-mail lists. How do we bring the communal support of the church play, with parents on the side whispering the lines to chirrens just to get them the experience, or the fact that we're always giving even mediocre speakers applause in a show of support, to digital spaces? What do those practices look like online?

These four related categories of research projects are only the merest of beginnings. Technology history and access can be so central to the study of African American rhetoric not only because of how Tyrone Taborn reminds us that it is a major site of struggle right now, or because that struggle is in so many ways, one over language, but because although cyberspace is constructed as White by default, so has every other kind of communication technology in our history, even if it is easy to forget that radio or print literacies or television is a technology. And a central part of African American rhetorical history has been about the battle over equal access to those technologies, and a concerted effort to use them to recover and celebrate African American history, and connect African Americans in resistance to racism. We have always been technology innovators and rhetorical innovators. We have always demanded equal access to technologies, without ever being satisfied in access for its own sake, demanding instead that those technologies be transformed to serve everyone more justly and equitably. Attention to this link between technologies and rhetorical production can help those interested in African American rhetoric and struggle to look forward to look back, to reflect on African American histories and futures in a highly technologized and highly racialized nation to reclaim and reimagine African American rhetorical study in a new century, to appreciate the ability we have always had, of discerning, in any given situation, all the available means and using them regardless of the barriers that still prevent equal access to them.

Finally, there is much that this book does not accomplish, many important experiences it does not document, many important ideas that it does not explore. This book is much like one of the quilts that slaves used to guide brave souls to freedom. It is one gathering of the scraps and bits around us and one attempt to make a map of them. It is one call for us all to pack our bags and find our own way through the laws, policies, brutality, rupture, networks of friends, systems of knowledge that can lead us to that higher, freer, ground. No matter how useful, or even beautiful in its usefulness, that one quilt cannot substitute for the journey. We have to be willing to get lost together, fight unfair laws and policies together, design and build new spaces together, document our stories together, play, fight, struggle, and celebrate together. Blackfolk have to keep underground traditions alive *and* participate fully in spaces that have excluded us, as do other groups of people of color. Those White institutions, disciplines, industries that have fostered the exclusions we face have to take responsibility for struggling for your own answers to the legacies of racism instead of expecting a diversity workshop or a guest speaker or sensitivity training or general conversations about tolerance and multiculturalism to solve correct those legacies. They must reach out to those who have been excluded, not as tokenized ways that ask them to speak for the race, but as central to collective work on the problems and potentials our technologies present us all with.

For those *still* wondering where to begin that search for higher ground, still wondering that they don't know enough about technology, about African American rhetoric, about scholarship, about teaching, about activism, spoken word artist Kamau Daood points the way in a line in his poetic sermon "Art Blakey's Drumsticks:" "those who carry themselves carry others until they can carry themselves."

ONE …

References

AfroGeeks. Retrieved from http://research.ucsb.edu/cbs/projects/afrogeeks.html. Accessed October 15, 2004.

Alkalimat, A. (2001a). "Technological revolution and prospects for Black liberation in the 21st century." *cy.rev* World Wide Web. Accessed June 13, 2001 from http://www.cyrev.net

Alkalimat, A. (2001b). "eBlack: A 21st Century Challenge." *eBlackStudies*. World Wide Web. Accessed June 13, 2001 from http://eblackstudies.net/eblack.html

Aristotle. (1956). *The politics*. E. Barker (Ed.). London: Oxford.

Asante, M. K. (1969). *Rhetoric of black revolution*. Allyn & Bacon.

Baraka, A. (1972). *Blues People*. New York: Quill.

Bell, D. (1989). *And we are not saved: The elusive quest for racial justice*. New York: Basic Books.

Blair, K. (1998). Literacy, dialogue, and difference in the "electronic contact zone." *Computers and Composition, 15*, 317–329.

Bolter, J. D. (1991). *Writing space: The computer, hypertext, and the history of writing*. Hillsdale, NJ: Lawrence Erlbaum Associates.

Bowers, D. L. (1996). When outsiders encounter insiders in speaking: Oppressed collectives on the defensive. *Journal of Black Studies, 26*(4), 490–503.

Brodie, H., & Graves, R. (1998). Masters, slaves, and infant mortality. *Technical Communication, 7*(4), 389–411.

Collier-Thomas, B. (1998). *Daughter of thunder: Black women preachers and their sermons*. San Francisco: Jossey-Bass.

Cone, J. (1995). *Martin and Malcolm and America: A dream or a nightmare?* Maryknoll, NY: Orbis Books.

Corsini, V., & Fogliasso, C. (1997). A descriptive study of the use of the Black communication style by African-Americans within an organization. *Journal of Technical Writing and Communication, 27*(1), 33–47.

Crawford, C. (Ed.). (2001). *Ebonics and language education*. New York: Sankofa World Press.

Crenshaw, C. (1997). Resisting Whiteness' rhetorical silence. *Western Journal of Communication, 61*(3), 253–278.

Cruse, H. (1984). *The crisis of the Negro intellectual.* New York: Quill.
Davis, A. (2002). Masked racism: Reflections on the prison industrial complex. In H. Boyd (Ed.), *Race and resistance: African Americans in the 21st century* (pp. 53–60). Cambridge, MA: South End Press.
Delgado, R., & Stefancic, J. (Eds.). (2000). *Critical race theory: The cutting edge.* Philadelphia, PA: Temple University Press.
Doheny-Farina, S. (1996). *The wired neighborhood.* New Haven, CT: Yale University Press.
DuBois, W. E. B. (1990). *The souls of Black folk.* New York: Library of America.
Dunbar-Nelson, A. M. (2000). *Masterpieces of Negro eloquence.* Mineola, NY: Dover Publications.
Early, G. (Ed.). (1992). *Speech and power: The Afro American essay and its cultural contents from polemics to pulpit: Vol. 1.* Hopewell, NJ: The Ecco Press.
Early, G. (Ed.). (1993). *Speech and power: The Afro American essay and its cultural contents from polemics to pulpit: Vol. 2.* Hopewell, NJ: The Ecco Press.
Eliot, T. S. (1922). "The Waste Land." In M. H. Abrams et al. (Eds.), *The Norton anthology of English literature, Vol. 2* (2003, pp. 268–283). NY: W.W. Norton & Company.
Falling Through the Net: A Survey of the "Have Nots" in Rural and Urban America. NTIA July 1995 http://www.ntia.doc.gov/ntiahome/fallingthru.html
Falling Through the Net II: New Data on the Digital Divide. NTIA July 1998 http://www.ntia.doc.gov/ntiahome/net2/
Falling Through the Net III: Defining the Digital Divide. NTIA July 1999 http://www.ntia.doc.gov/ntiahome/fttn99/contents.html
Falling Through the Net IV: Toward Digital Inclusion. NTIA July 2000 http://www.ntia.doc.gov/ntiahome/fttn00/contents00.html
Feenberg, A. (1991). *Critical Theory of Technology.* Oxford: Oxford University Press.
Feenberg, A. (1995a). *Alternative modernity: The technical turn in philosophy and social theory.* Berkeley, CA: University of California.
Feenberg, A. (1995b). Subversive rationalization: Technology, power, and democracy. In A. Feenberg & A. Hannay (Eds.), *Technology and the politics of knowledge.* (pp. 3–23) Bloomington: Indiana University Press.
Foner, P. S. (1975). *Voice of Black America.* Capricorn Books.
Foner, P. S., & Branhams, R. J. (1998). *Lift every voice.* University of Alabama Press.
Franklin, V. P., & Collier-Thomas, B. (1990). Biography, race vindication, and African American intellectuals: Introductory essay. *Journal of Negro History, 81*(1), 1–16.
Galster, G. C., & Hill, E. W. (Eds.). (1992). *The metropolis in Black and White: Place, power and polarization.* Rutgers, NJ: Rutgers University Press.
Gandy, O. (1995). It's discrimination, stupid! In J. Brook & I. Boal (Eds.), *Resisting the virtual life: The culture and politics of information* (pp. 35–47). San Francisco: City Lights.
Garvey, M. (1923). An exposé of the caste system among Negroes. In M. Marable & L. Mullings (Eds.), *Let nobody turn us around: Voices of reform, resistance, and renewal.* Lanham, MD: Rowan and Littlefield.
Gates, H. L. (1986). *Race, writing, and difference.* Chicago: University of Chicago.

Gates, H., & McKay, N. (1997). *The Norton anthology of African American literature* (pp. xxxvii–xli). New York: W.W. Norton & Company.

Gillette, H. (1995). *Between justice and beauty: Race, planning, and the failure of urban policy in Washington, D.C.* Baltimore: Johns Hopkins University Press.

Gilyard, K. (1991). *Voices of the self: A study in language competence.* Detroit, MI: Wayne State University Press.

Goldsmith, W. W., & Blakely, E. J. (Eds.). (1992). *Separate societies: Poverty and inequality in U.S. cities.* Philadelphia, PA: Temple University Press.

Grabill, J. T. (1998). Utopic visions, the technopoor, and public access: Writing technologies in a community literacy program. *Computers and Composition, 15,* 297–315.

Harris, M. D. (2003). *Colored pictures: Race and representation.* Chapel Hill: University of North Carolina Press.

Heidegger, M. (1986). *The question concerning technology and other essays.* New York: Harper.

Hesse, D. (1999). Saving a place for essayistic literacy. In. G. Hawisher & C. Selfe (Eds.), *Passions, pedagogies, and 21st century technologies* (pp. 34–48). Logan: Utah State University Press.

Higginbotham, A. L. (1998). *Shades of freedom: Racial politics and the presumptions of the American legal process.* New York: Oxford.

Hoffman, D., & Novak, T. (1998). *Bridging the digital divide: The impact of race on computer access and computer use.* Accessed December 5, 2001 from wwww2000.ogsm.vanderbilt.edu/papers/race/science.html

Holmes, D. G. (1999). Fighting back by writing Black: Beyond racially reductive composition theory. In. K. Gilyard (Ed.), *Race, rhetoric, and composition* (pp. 53–56). Portsmouth, NH: Boynton/Cook.

Hughes, L. (1998). The Negro artist and the racial mountain. In P. L. Hill (Ed.), *Call and response: The riverside anthology of African American literature* (p. 1267). New York: Riverside.

Jafa, A., & Tate, G. (1998). From dogon to digital: Design force 2000. Looting other disciplines along the way. *International Review of African American Art, 13*(1).

Johnson, R. (1998). *User centered technology: A rhetorical theory for computers and other mundane artifacts.* Albany: State University of New York.

Johnson-Eilola, J. (1997). *Nostalgic angels: Rearticulating hypertext writing.* Norwood, NJ: Ablex.

Joyce, M. (1995). *Of two minds: Hypertext pedagogy and poetics.* Ann Arbor, MI: Michigan.

Kelley, R. D. G. (2002). *Freedom dreams: The Black radical imagination.* Boston: Beacon Press.

King, M. L. (1991). Remaining awake through a great revolution. In M. Washington (Ed.), *Testament of hope: The essential writings and speeches of Martin Luther King, Jr.* (pp. 197–201). San Francisco: Harper.

Kolko, B. E. (2000). Erasing @race: Going White in the (inter)face. In B. E. Kolko, L. Makamura, & G. B. Rodman (Eds.), *Race in cyberspace* (pp. 213–232). New York: Routledge.

Kreuzer, T. (1993). Computers on campus: The Black-White technology gap. *Journal of Blacks in Higher Education, 12*(4), 88–95.

Lokko, L. (Ed.). (2002). *White papers, Black marks: Architecture, race, culture*. Minneapolis, MN: University of Minnesota Press.

MacGillis, A. (2004, September 19–25). Poor schools, rich targets. Retrieved October 15, 2004 from the *Baltimore Sun* http://www.baltimoresun.com/nationworld/balte.software21sep21,1,1793980.story?ctrack=1&cstet=true

Malcolm X. (1965). The ballot or the bullet. In G. Breitman (Ed.), *Malcolm X speaks: Selected speeches and statements* (pp. 23–44). New York: Pathfinder.

Malveaux, J. (2004). Extended quote from Julianne Malveaux—Where? When? Amy Alexander. "Having Our Say." Accessed November 1, 2004 from http://www.africana.com

Marable, M. (2002). *The great wells of democracy: The meaning of race in American life*. New York: BasicCivitas Books.

Marable, M., & Mullings, L. (Eds.). (2000). *Let nobody turn us around: Voices of reform, resistance, and renewal*. Lanham, MD: Rowan and Littlefield.

Mitchell, M. (2001). *The crisis of the African American architect: Conflicting cultures of architecture and (Black) power*. San Jose, CA: Writer's Club Press.

Mitchell, W. J. (1995). *City of bits: Space, place, and the infobahn*. Cambridge, MA: Massachusetts Institute of Technology Press.

Mitchell, W. J., Schon, D. A., & Mitchell, W. J., (Eds.). (1999). *High technology and low income communities: Prospects for the positive use of advanced information technology*. Cambridge, MA: Massachusetts Institute of Technology Press.

Moran, C. (1999). Access: The a-word in technology studies. In G. E. Hawisher & C. L. Selfe (Eds.), *Passions, pedagogies, and 21st century technologies* (pp. 205–220). Logan, UT: National Council of Teachers of English.

Moran, C., & Selfe, C. L. (1999). Teaching English across the technology/wealth gap. *English Journal, 88*(6), 48–55.

Morrison, T. (1988). *Beloved*. New York: Penguin.

Moses, W. J. (1978). *The golden age of Black nationalism, 1850–1920*. New York: Oxford.

Moss, B. (2003). *A community text arises: A literate text and a literacy tradition in African American churches*. Creskill, NJ: Hampton Press.

Nakamura, L. (2000). Where do you want to go today? In B. E. Kako, L. Nakamara, & G. B. Rodman (Eds.), *Race in Cyberspace* (pp. 15–23). New York: Routledge.

Neal, L. (1999). The Black aesthetic. In H. A. Ervin (Ed.), *African American literary criticism* (pp. 122–128). New York: Twayne.

Nelson, A. (2001). *Technicolor*. New York: New York University Press.

Ong, W. (1982). *Orality and literacy*. London: Routledge.

Pendergrast, C. (1998). Race: The present absence in composition studies. *College Composition and Communication, 50*(1), 36–53.

Pitney, D. H. (1990). *The Afro American jeremiad: Appeals for justice in America*. Philadelphia, PA: Temple University Press.

Porter, J. E. (1998). *Rhetorical ethics and internetworked writing*. Greenwich, CT: Ablex.

Powell, L. (1990). Factors associated with the underrepresentation of African Americans in mathematics and science. *Journal of Negro Education, 59*(3), 292–298.

Rawlinson, R. (1999). Can Robert Johnson bring more Blacks online? Accessed December 1, 2001 from http://archive.salon.com/tech/feature/1999/10/06/bet_johnson/print.html

Richardson, E. (1997). African American women instructors: In a net. *Computers and Composition, 14*(2), 279–287.

Richardson, E. (2003). *African American literacies.* London: Routledge.

Rickford, J. R., & John, R. (2000). *Spoken soul: The story of Black English.* New York: Wiley.

Romano, S. (1993). The egalitarianism narrative: Whose story? Whose yardstick? *Computers and Composition, 10*(3), 7–28.

Rose, T. (1991). Fear of a Black planet: Rap music and Black cultural politics in the 1990s. *Journal of Negro Education, 60*(3), 276–290.

Rose, T. (1994). *Black noise: Rap music and Black culture in contemporary America.* Hanover, NH: Wesleyan University Press.

Selber, S., & Johnson-Eilola, J. (2001). Sketching a framework for graduate education in technical communication. *Technical Communication Quarterly, 10,* 403–437.

Selfe, C. L. (1999). *Technology and literacy in the twenty-first century: The importance of paying attention.* Carbondale, IL: Southern Illinois University Press.

Siedler, H. (1978). *Planning and building down under: New settlement strategy and architectural practice in Australia.* Vancouver, BC: University of British Columbia Press.

Slater, R. (1994). Will Blacks in higher education be detoured off the information superhighway? *Journal of Blacks in Higher Education, 3,* 96–99.

Smitherman, G. (1986). *Talkin and testifyin: The language of Black America.* Detroit, MI: Wayne State University Press.

Smitherman, G. (2000). *Talkin that talk: Language, culture, and education in African America.* New York: Routledge.

Stull, B. (1999). *Amid the fall, dreaming of Eden: DuBois, King, Malcolm X and emancipatory composition.* Carbondale, IL: Southern Illinois University Press.

Sullivan, P., & Dauterman, J. (1996). Issues of written literacy and electronic literacy in workplace settings. In P. Sullivan & J. Dauternman (Eds.), *Electronic literacies in the workplace* (pp. vii–xxxiii). Urbana, IL: National Council of Teachers of English.

Taborn, T. (2001). The art of Tricknology. *U.S. Black engineer: Information technology.* Baltimore, MD: Career Communication Group.

Tal, K. (1996, October). The unbearable Whiteness of being: African American critical theory and cyberculture. *Wired Magazine, 4*(10).

Taylor, C. A. (1996). *Defining science: A rhetoric of demarcation.* Madison, WI: University of Wisconsin.

Tobin, J. L., & Dobard, R. (2000). *Hidden in plain view: A secret story of quilts and the underground railroad.* New York: Anchor.

Toomer, J. (1988). *Cane.* New York: Norton.

Turner, P. (1993). *I heard it through the grapevine: Rumor in African American culture.* Los Angeles: University of California.

Vielstemmig, M. (1999). Petals on a wet, black bough. In G. Hawisher & C. Selfe (Eds.). *Passions, pedagogies, and 21st century technologies.* Logan: Utah State University Press.

Villaneuva, V. (1993). *Bootstraps: From an American academic of color.* Urbana, IL: National Council of Teachers of English.

Walker, D. (1830). Appeal in four articles: Together with a preamble, to the coloured citizens of the world, but in particular and very expressly, to those of the United States of America. In. M. Marable & L. Mullings (Eds.), *Let nobody turn us around: Voices of reform, resistance, and renewal*. Lanham, MD: Rowan and Littlefield.

Walker, J. (1994). The body of persuasion: A theory of the enthymeme. *College English, 56*(1), 46–63.

Wilkins, C. (1998). A style that nobody can deal with: Notes from the Doo Bop Hip Hop Inn. *International Review of African American Art.*

Winner, L. (1986). *The whale and the reactor: A search for limits in an age of high technology*. Chicago: University of Chicago.

Woodson, C. G. (1969) *Negro orators and their orations*. New York: Russell & Russell.

Wright, G. (2000). Course Syllabus. Columbia University. World Wide Web. Accessed April 15, 2000 from http://www.arch.columbia.edu/Admin/Syllabi/24529.html

Wysocki, A., & Johnson-Eiola, J. (1999). Blinded by the letter: Why are we using literacy as a metaphor for everything else? In G. Hawisher & C. Selfe (Eds.), *Passion, pedagogies, and 21st century technologies*. Logan, Utah State University Press.

Young, H. A., & Young, B. H. (1977). Science and Black studies. *Journal of Negro Education, 46*(4), 380–387.

Author Index

J

K

L

M

N

O

P

R

S

T

V

W

Y

Subject Index

M

N

O